Life Science "Do-Nows" and "Exit Tickets"

180 Days of Warm-Up and Closure Activities

Written and Edited by G. Katz Chronicle

Publisher: IMNF Education Press

IMNF Education Press grants teachers permission to photocopy and/or electronically download and print pages from this book for classroom use. No other part of this publication may be reproduced in whole or in part, or stored in a retrieval system, or transmitted in any form or by any means, electronic, mechanical, photocopying, recording, or otherwise without written permission of the publisher. For information regarding permission, write to IMNF Education Press at IMNFEducationPress@gmail.com.

Cover Design: G. Katz Chronicle

Copyright 2018

First edition

This work is dedicated to Gloria Closi, without whom this endeavor would've been a much more arduous process. Many thanks!

Table of Contents

Introduction 1

Biochemistry 2

    Life's First Molecule Was Protein, Not RNA, New Model Suggests (1) 3

    Boost for Lipid Research: Researchers Facilitate Lipid Data Analysis (2) 5

    Chemists Synthesize Millions of Proteins Not Found in Nature (3) 7

    Can Diet Delay Menopause? (4) 9

    Dining Out Associated with Increased Exposure to Harmful Chemicals (5) 11

    Don't Waste Your Money on DNA-Based Diets (6) 13

    Do Vitamin Supplements Improve Your Health? (7) 15

    Could Eating More Seafood Help Couples Conceive? (8) 17

    'Lipid Asymmetry' Plays Key Role in Activating Immune Cells (9) 19

    Researchers Defy Biology: Mice Remain Slim on Burger Diet (10) 21

    New Strains of Staple Crops Serve Up Essential Vitamins (11) 23

    Can Eating Pasta Really Help You Lose Weight? (12) 25

    Some Vitamins and Minerals May Carry More Risks Than Benefits (13) 27

Cellular Energy 29

    Could Scientists Have Found Multicellular Animals
        That Live in Oxygen-Free Environment? (1) 30

    Artificial 'Plants' Could Fuel the Future (2) 32

    Plants Use Light Trick to Make Bees See Blue (3) 34

    Corpse Flower (4) 36

    Do Fish Get Out of Breath? (5) 38

Engineers Create Plants That Glow (6) 40

Future Increase in Plant Photosynthesis Revealed by
    Seasonal Carbon Dioxide Cycle (7) 42

Mistletoe Lacks Key Energy-Generating Complex (8)..44

Mitochondria Have Their Own DNA (9) 46

Mitochondria: Cell Powerhouses and More (10)..48

Photosynthesis More Ancient Than Thought,
    and Most Living Things Could Do It (11)..50

Key Protein in Cellular Respiration Discovered (12)...52

The Algae Invasion (13) 54

Does Training at High Altitudes Help Olympians Win? (14)...56

Vitamin B5 (15)...58

Classification 60

4 New Legless Lizards Discovered in California (1)..61

46 New Creatures Found in Rainforest (2) 63

Breeding Benefits When Love Bites Wombats on The Butt (3) 65

Classifying Algae (4)...67

Mathematics Supports a New Way to Classify Viruses Based on Structure (5) 69

Differences Between Reptiles and Amphibians (6) 71

Two New Dog-Faced Bat Species Discovered (7)...73

From Aristotle to Linnaeus: The History of Taxonomy (8) 75

How Are Animals Classified? (9) 77

How Are Plants Classified? (10) 79

Nature's Misfits: Reclassifying Protists Helps Answer
    How Many Species Remain Undiscovered (11) 81

Biologists May Have Found a New Fungal Group (12) 83

What do Piranhas and Goldfish Have in Common? (13) 85

Platypus: The Unique Mammal Who Might Be Able to Fight Bacteria (14) 87

A Few Bad Scientists Are Threatening to Topple Taxonomy (15) 89

Ecology 91

Beavers Could Help Curb Soil Erosion, Clean Up Polluted Rivers (1) 92

Old Maps Highlight New Understanding of Coral Reef Loss (2) 94

What is the Environmental Impact of Your Sandwich? (3) 95

The Terrifying Way Fire Ants Take Advantage of Hurricane Floods (4) 98

If People Were Cockroaches, Adapting to Climate Change Would Be Easy (5) 100

Inside Australia's War on Invasive Species (6) 102

For Lemurs, Size of Forest Fragments May Be
    More Important Than Degree of Isolation (7) 104

Lichens Are Bioindicators (8) 106

Lichens Can Be Made of Three Organisms, Not Just Two (9) 108

Meet the Lichens: Symbionts and Pioneers (10) 110

Taking Manatees Off the Endangered Species List Doesn't
    Mean We Should Stop Protecting Them (11) 112

Migratory Animals Carry More Parasites (12) 114

Rabbit Relatives Reel from Climate Change (13) 116

Resurrecting Extinct Animals Might Do More Harm Than Good (14) 118

Russian Cuckoos Are Taking Over Alaska (15) 120

Sensors Applied to Plant Leaves Warn of Water Shortage (16) 122

Springtime Now Arrives Earlier for Birds (17) 124

Tick Paralysis (18) 126

Evolution 128

How Did 3D Vision Develop? (1) 129

The Battle of the Sexes Can Show Us How to Live Longer (2) 131

Are Cities Affecting Evolution? (3) 133

When Dinosaurs Took Flight (4) 135

**Earth's Orbital Changes Have Influenced Climate,
Life Forms For At Least 215 Million Years (5)** 137

The First Dogs May Have Been Domesticated in Central Asia (6) 139

How Fish Conquered the Land (7) 141

How Language Evolved (8) 143

Humans Don't Use More Brainpower Than All Other Animals (9) 145

**New Species Can Develop in as Little as Two Generations,
Galapagos Study Finds (10)** 147

Nose Shape Dependent on Ancestral Climate (11) 149

**Archaeologists May Have Discovered One of the
Earliest Examples of a 'Crayon' (12)** 151

Pigeons Can Read a Little Bit, New Research Shows (13) 153

Prehistoric Women had Stronger Arms Than Today's Elite Rowing Crews (14) 155

The Rise of the Mammals (15) 157

Scientists Discover a Distant Relative of Today's Horses (16) 159

Amber Discovery Indicates Lyme Disease is Older Than Human Race (17) 161

Sixteenth Century Turkey Bones Found Under a Street in England (18) 163

Venomous Slow Loris May Have Evolved to Mimic Cobras (19) 165

Genetics 167

Abominable Snowman? Nope: Study Ties DNA Samples from
Purported Yetis to Asian Bears (1) 168

Why Some Cats Look Like They Are Wearing Tuxedos (2) 170

First Primate Clones Produced Using the "Dolly" Method (3) 172

Cold Snap Shapes Lizard Survivors (4) 174

Doctors Fix Genetic Defect and Grow Boy New Skin (5) 176

You May Be as Friendly as Your Genes (6) 178

Researchers Find Genetic 'Dial' Can Control Body Size in Pigs (7) 180

Can Gene Editing Save the World's Chocolate? (8) 182

Gene Therapy Restores Sight to Blind Mice (9) 184

Gray Hair Linked to Immune System Activity and Viral Infection (10) 186

The Happy Hour Gene (11) 188

Is "Junk DNA" What Makes Humans Unique? (12) 190

Link Between Biological Clock and Aging Revealed (13) 192

New Way to Activate Stem Cells to Make Hair Grow (14) 194

Tree Rings Used to Counter Smugglers' Rings (15) 196

Simple Urine Test Could Measure How Much Our Body Has Aged (16) 198

Where do heart cells come from? (17) 200

Human Body Systems 202

Autism Symptoms Improve After Fecal Transplant, Small Study Finds (1) 203

Many Babies Healthier in Homes with Dogs (2) 205

A Better Flu Shot May Be on the Horizon (3) 207

Body Hair Is Natural (4) 209

Celebrate Valentine's Day by Eating an Actual Heart (5) 211

**Take a Deep Breath...Hold it... and Let Go...
    You Should Feel Better Now- - Science Says So! (6)** 213

Dehydration in the Human Body (7) 215

Are e-Cigarettes and Vaping Safer Than Traditional Cigarettes? (8) 217

The Effect of Age on Eyesight (9) 219

Energy from Pee (10) 221

Is Expired Sunscreen Better Than No Sunscreen? (11) 223

**Comparing Food Allergies: Animals and Humans
    May Have More in Common Than You Think (12)** 225

When HIV Drugs Don't Cooperate (13) 227

Local Honey Might Help Your Allergies—But Only If You Believe (14) 229

How to Tell if You Really Have a Fever (15) 231

Human-Dog Bond Provides Clue to Treating Social Disorders (16) 233

Being Hungry Shuts Off Perception of Chronic Pain (17) 235

What's the Difference Between Indoor and Outdoor Allergies? (18) 237

Common Ingredient in Packaged Food May Trigger Inflammatory Disease (19) 239

Students Know About Learning Strategies -- But Don't Use Them (20) 241

**Regular Physical Activity is 'Magic Bullet' For
    Pandemics of Obesity, Cardiovascular Disease (21)** 243

Our Muscles Measure the Time of Day (22) 245

Tissue-Engineered Human Pancreatic Cells Successfully Treat Diabetic Mice (23) 247

Six Genes Driving Peanut Allergy Reactions Identified (24) 249

What the Consistency of Your Poo Says About Your Health (25) 251

A New Reason Why Newborns Can't Fight Colds (26) 253

Here's What Happens to Your Body If You Eat Too Much Salt (27) 255

Self-Esteem in Kids: Lavish Praise Is Not the Answer, Warmth Is (28) 257

Smartphone-Controlled Smart Bandage for Better, Faster Healing (29) 259

Does Your Back Feel Stiff? Well, It May Not Actually Be Stiff, Study Finds (30) 261

Is It Possible to Eat So Much That Your Stomach Explodes? (31) 263

Increased Exposure to Sunlight May Be Good for Some People (32) 265

Veterinary Surgeons Perform First-Known Brain Surgery to
   Treat Hydrocephalus in Fur Seal (33) 267

Why Are Suicide Rates Rising? (34) 269

You Are Disgusting in So Many Ways (35) 271

Answer to Young People's Persistent Sleep Problems (36) 273

Reproduction & Development 275

New Gene May Explain Why Some Men More Likely to Have Sons (1) 276

Our Mothers' Exposure to BPA Might Lead us to Binge Eat as Adults (2) 278

Scientists are Puzzling Out How Butterflies
   Assemble Their Brightly Colored Scales (3) 280

Scientists Discover the Embryonic Origins of the Phallus (Genital Organs) (4) 282

Cells Driving Gecko's Ability to Re-grow its Tail Identified (5) 284

Heteropaternal Superfecundation: Who's Your Daddy? (6) 286

Armadillo, Hedgehog and Rabbit Genes Reveal How Pregnancy Evolved (7) 288

Can We Stay Young Forever, Or Even Recapture Lost Youth? (8) 290

How Mum's Immune Cells Help Her Baby Grow (9) 292

An Old Drug for Alcoholism Finds New Life as Cancer Treatment (10) 294

Self-Fertilizing Fish Have Surprising Amount of Genetic Diversity (11) 296

Sex Reversal in Bearded Dragons Creates Females That Behave Like Males (12) 298

Skin Cells Remember Previous Injury (13) 300

Scientific Inquiry 302

Behavior in High School Predicts Income and Occupational Success Later in Life (1) ... 303

Caffeine's Sport Performance Advantage for
Infrequent Tea and Coffee Drinkers (2) 305

Cat Poop Parasites Don't Make You Psychotic (3) 307

Dinosaur Parasites Trapped in 100-Million-Year-Old Amber
Tell Blood-Sucking Story (4) 309

Does Dim Light Make Us Dumber? (5) 311

Dogs Are More Expressive When Someone Is Looking (6) 313

Driving Speed Affected When a Driver's Mind "Wanders" (7) 315

NIH Ends All Research on Chimps (8) 317

Even Insects Have Distinct Personalities (9) 319

Freezing Hunger-Signaling Nerve May Help Ignite Weight Loss (10) 321

Bonobos Prefer Hinderers over Helpers (11) 323

No Evidence to Support Link Between Violent Video Games and Behavior (12) 325

Music Taste Linked to Brain Type (13) 327

Our Reactions to Odor Reveal Our Political Attitudes, Survey Suggests (14) 329

Poor Grades Tied to Class Times That Don't Match Our Biological Clocks (15) 331

    This Is Why Science Loves Twins (16) 333

    New Study Finds That More Screen Time Coincides with
        Less Happiness in Youths (17) 335

    Sorry, Grumpy Cat: Study Finds Dogs are Brainier Than Cats (18) 337

Study of Life 339

    Are Viruses Alive? New Evidence Says Yes (1) 340

    Understanding Bacterial Wargames Inside Our Body (2) 342

    Bacteria on Kitchen Towels (3) 344

    Rabbit, Dog, Human: How One Bacterial Infection Spread (4) 346

    Cellular Messengers Communicate with Bacteria in the Mouth (5) 348

    Fecal Transplantation (6) 350

    Good Germs Can Be Found in Poop (7) 352

    How Often Should I Clean My Phone? (8) 354

    Mummies' Tummies to Reveal Digestive Evolution (9) 356

    New Antibiotic Found in Human Nose (10) 358

    A 508-Million-Year-Old Sea Predator With a 'Jackknife' Head (11) 360

    Peptidoglycan: The Bacterial Wonder Wall (12) 362

    If You have Pimples You May Need Better Bacteria (13) 364

    Swimming Pools Are Full of Poop, But They Probably Won't Make You Sick (14) 366

    Powered by Poop (15) 368

    Probiotics Might Help Allergies, But We're Not Sure How (16) 370

Suggested Responses to Fill-Ins 372

Sample Exit Tickets 387

Bibliography 388

Request for a Review 401

Connect with The Author 402

Introduction

Like many teachers, I want my students engaged in purposeful learning as soon as they enter my classroom. Toward that end, many of my colleagues have used "do-now", "warm-up", or other similar introductory activities to focus students during the first few minutes of class as they take attendance or do other classroom "housekeeping" chores.

I have struggled to find appropriate class starter activities for the learners who pass through my door because of their divergent ability levels. Of course, the ideal starter activity should be thought provoking and somewhat challenging, but not so difficult that students are filled with questions about how to complete the activity.

This book represents a compilation of science articles which I have abbreviated or summarized to the confines of one page. Within each article, blanks have replaced ten words, which can be inferred through the context of the article. The uniformity of this format removes student bewilderment as to how to complete the activity.

If practiced daily from the start of the school year, students learn that they merely are to make their best educated guess at the missing words. Generally, the missing words are repeated throughout the article. Of course, practitioners may decide to provide students a word bank, if they choose. The answers provided at the back of this text are suggested, as alternate words may also make sense in context.

The back of each warm-up is a blank lined page. I have designed this format for student use as an "exit-ticket" to summarize the day's learning. A series of exit-ticket starter statements have been provided. Two to three statements can be projected at the end of class for students to copy and complete. These statements can be rotated throughout the school year.

Although I compiled these 180 articles for use as do-nows in my biology classroom, the articles could be used during general classroom instruction and discussion. Many of the articles would also be appropriate for health classrooms. Ultimately, the teacher knows their students best and will determine how and if this instructional tool can help their students.

Respectfully Offered,

G. Katz Chronicle

Biochemistry

Life's First Molecule Was Protein, Not RNA, New Model Suggests (1)

Directions: Fill in a word that best completes each sentence.

Basic chemical building blocks, which existed on early earth, likely grew into longer macromolecules (polymers) that could self-replicate. These **1**_____-replicating polymers could perform functions essential to life: They could store information and catalyze chemical reactions. Scientists who study the origin of life have long wondered whether these basic building blocks were proteins or nucleic acids.

For most of the history of life, **2**_____ acids (DNA and RNA) stored information, the instructions for making proteins; proteins have extracted and copied the instructions from the nucleic **3**_____, as well as catalyzed chemical reactions. Scientists believe one of those molecules, either the nucleic acids or the **4**_____, originally handled the storing of information *and* the catalyzing of chemical **5**_____.

The "RNA world" hypothesis proports that life emerged out of RNA, which could replicate itself. RNA can store information, fold, and catalyze reactions, like proteins do. However, because RNA is complex and sensitive, some experts are skeptical that it could have arisen spontaneously under the harsh conditions of the prebiotic world. Moreover, both RNA molecules and proteins must take the form of long, folded chains to do their catalytic work, and early earth's harsh environment would likely have prevented either nucleic acids or amino acid chains from getting long enough.

A possible solution to the question of which developed first, proteins or amino acids, has been proposed by researchers Ken Dill and Elizaveta Guseva of Stony Brook University and Ronald Zuckermann of the Lawrence Berkeley National Laboratory. The scientists presented a hydrophobic-polar (HP) protein folding model to the Proceedings of the National Academy of Sciences (PNAS). The **6**_____-polar (HP) protein-folding model treats the 20 amino acids as just two types of subunit: blue, water-loving beads (polar monomers) and red, water-hating beads (nonpolar monomers). The model can fold a chain of these beads in sequential order. The arrangement depends on the tendency for the red, hydrophobic beads to clump together so that they can better avoid **7**_____. Together, they act as a catalyst for elongating polymers, speeding up reactions. Once folded, the molecules can continue to form long polymers in greater and greater numbers.

Andrew Pohorille, of NASA, believes proteins likely came first because they are easier to make than nucleic acids. However, before this model, scientists couldn't answer how replication of proteins occurred. "This is an attempt to show that even though you cannot really **8**_____ proteins the same way you can replicate RNA, you can still build and evolve a world without that kind of precise information storage."

The model still needs to be tested against other theoretical models in the lab if it is to replace the "**9**_____ world" hypothesis. The HP model is highly simplified and doesn't account for many of the complicated molecular details and chemical interactions that characterize real life. For now, the "RNA world" **10**_____ is more widely accepted. Still, the scientists are optimistic about further research.

Boost for Lipid Research: Researchers Facilitate Lipid Data Analysis (2)

Directions: Fill in a word that best completes each sentence.

Lipids form cell walls, store energy and release it when necessary, and play an important role in cell signaling in all living things. They are necessary for life. Changes in the composition of lipids can cause illnesses such as cancer, fatty liver, and multiple sclerosis. Therefore, scientists are interested in learning how to tell the types of **1**_____ apart and notice changes in them to aid in the detection of disease.

To detect lipids which are indicators for **2**_____, healthy and sick organisms are compared quantitatively. This **3**_____ requires reliable and detailed information about the structure and composition of lipids from tissue samples. Researchers from the BioTechMed-Graz initiative have developed a tool to measure lipids which is presented *Nature Methods*.

Lipids (fats) are complex substances, which, in addition to various other components, predominantly consist of fatty acids. There are still many unknowns, however, about the structural properties of lipid molecules. An analysis process called high-throughput profiling is still in its infancy with regard to lipid study.

BioTechMed-**4**_____ researchers used a **5**_____-throughput method whereby many samples of lipids are measured using mass spectrometry. The data, called spectra, provide information for identification of the type and class of lipid or the type and position of fatty acyl chains. Up until now there has not been a universal system for automating the detection of lipid structures because there is such a great spectral diversity among them. Even the same types of **6**_____ can show different fragments in the **7**_____ depending on the setup of the mass spectrometer and the ionization.

Researchers involved in the study explain that, "fast and reliable details on the lipid composition of cell samples are a prerequisite for comparisons with reference samples from healthy cells -- which are required for the detection of biomarkers characteristic for diseases." Researchers want to figure out which changes in the lipid **8**_____ of cells are relevant in diagnostics.

The Lipid Data Analyzer will facilitate work in biomedical research. Jürgen Hartler, of the Institute of Computational Biotechnology, says, "The method which we have developed in collaboration with colleagues from Med Uni Graz and Uni Graz, interprets lipid spectra using intuitive rule sets and can be flexibly accommodated to various fragmentation characteristics. This makes it possible for the first time to identify lipids at a very detailed structural level more precisely and reliably than previous solutions."

The Lipid Data **9**_____ detected more than 100 novel lipids which were previously unreported. The tool can be flexibly adapted. It is not just for new classes of lipids. It may be used, for instance, to characterize polysaccharides and glycolipids, which are lipids with attached sugars. The researchers provided their Lipid **10**_____ Analyzer as an open resource to the scientific community.

Chemists Synthesize Millions of Proteins Not Found in Nature (3)

Directions: Fill in a word that best completes each sentence.

Proteins produced by living cells are made from 20 natural amino acids. Proteins function as enzymes in chemical reactions, antibodies in reacting to antigens, and regulate the activity of cells, as well as numerous other life functions. **1**_____ also play a role in movement, structural support, storage, communication between cells, digestion and the transport of substances around the body.

A group of MIT chemists recently came up with a way to synthesize (assemble) proteins from **2**_____ acids not used in nature, including many that are mirror images of natural amino **3**_____. These proteins, which the researchers call "xenoproteins," offer many advantages over naturally occurring proteins. They are more stable, meaning that unlike most protein drugs, they don't require refrigeration, and may not provoke an immune response. The presentation of these xenoproteins can be found in a paper appearing in the *Proceedings of the National Academy of Sciences*.

The research team built on technology that Brad Pentelute's lab had previously developed for rapidly synthesizing protein chains. His tabletop machine can perform all the **4**_____ reactions needed to string together amino acids, synthesizing the desired proteins within minutes. Before this study, no research group had been able to create so many proteins made purely of non-natural amino acids.

After synthesizing the **5**_____, the researchers screened them to identify proteins that would bind to an IgG antibody against an influenza virus surface protein. The antibodies were tagged with a fluorescent molecule and then mixed with the xenoproteins. Using a system called fluorescence-activated cell sorting, the researchers were able to isolate xenoproteins that bind to the fluorescent IgG molecule. This screen, which can be done in only a few hours, revealed several xenoproteins that **6**_____ to the target. In other experiments the researchers also identified xenoproteins that bind to anthrax toxin and to a glycoprotein produced by the Ebola virus.

The researchers are now working on synthesizing proteins modeled on different scaffold shapes, and they are searching for xenoproteins that bind to other potential drug targets. Their long-term goal is to use this system to rapidly **7**_____ and identify proteins that could be used to neutralize any type of emerging infectious **8**_____. "The hope is that we can discover molecules in a rapid manner using this platform, and we can chemically manufacture them on demand. And after we make them, they can be shipped all over the place without **9**_____, for use in the field," Pentelute says. In addition to potential drugs, the researchers also hope to develop "xenozymes" -- xenoproteins that can act as enzymes to catalyze novel types of chemical **10**_____.

Can Diet Delay Menopause? (4)

Directions: Fill in a word that best completes each sentence.

Eating certain foods may be linked to a delayed or hastened onset of menopause, a new study from England finds. Fresh legumes, such as peas and green beans and oily fish, such as salmon, sardines and mackerel may help to a delay the onset of **1**_____, while eating refined carbs, such as rice and pasta, may relate to an earlier onset, the researchers found. The study found only a correlation between diet and the timing of menopause. It did not look at what specific mechanisms would enable a woman's diet to influence menopause. In other words, the study didn't prove cause and **2**_____.

Additionally, it's not yet clear whether delaying menopause is a good thing. Early menopause is linked to an increased risk of heart disease, osteoporosis and depression, but it also protects against breast, endometrial and ovarian cancers. "As such, we cannot really recommend women to consume these specific foods to influence their **3**_____ of natural menopause," said lead study researcher Yashvee Dunneram, a doctoral student at the University of Leeds in England.

Previous studies have shown that **4**_____ may influence menopause, but results from different studies had contradictory findings, said Dunneram. That's why she decided to use survey data from the U.K. Women's Cohort Study, which allowed her and her colleagues to examine the eating habits of **5**_____ before they reached menopause and then compare that information with the women's actual age of menopause. This ongoing survey made the data more reliable than a retrospective study, in which people try to remember what they ate years afterward, she said.

To investigate, the researchers looked at survey **6**_____ collected from more than 35,000 women, ages 35 to 69, from England, Scotland and Wales. On the **7**_____, the women answered questions about health factors that might influence menopause, including diet, weight history, exercise levels, reproductive history and the use of hormone replacement therapy. Then, four years later, the **8**_____ followed up with the women and asked at what age the women had reached menopause. About 14,000 women responded at both time points, and of those, 914 reported that they had gone through natural menopause during a four-year period, when they were between ages 40 and 65. On average, the women reached menopause at **9**_____ 51, the researchers found.

Perhaps certain foods, such as grapes, legumes and oily fish, are associated with later menopause because they contain or stimulate antioxidants, which may play a role in egg maturation and release, the researchers said. Conversely, refined **10**_____ increase the risk of insulin resistance, which can interfere with sex hormone activity and boost estrogen levels. Those are factors that may increase the number of menstrual cycles and deplete egg supply faster, the researchers said.

Dining Out Associated with Increased Exposure to Harmful Chemicals (5)

Directions: Fill in a word that best completes each sentence.

Dining out more than eating home-cooked food may boost levels of potentially health-harming chemicals called phthalates in the body, according to a study. Phthalates, a group of chemicals used in food packaging and processing, are known to disrupt human hormones. People who reported eating out more often had phthalate levels 35 percent higher than those who reported eating home-cooked food. "This study suggests food prepared at home is less likely to contain high levels of **1**_____, chemicals linked to fertility problems, pregnancy complications and other health issues," says Ami Zota of George Washington University.

Lead author Julia Varshavsky, of the University of California, and Zota used data from the National Health and Nutrition Examination Survey (NHANES) collected between 2005 and 2014. The 10,253 survey participants were asked what they ate and where the food came from in the previous 24 hours. The links between what people **2**_____ and the levels of phthalate break-down products found in urine samples were then analyzed.

Most (61%) reported dining out the previous day. In addition, the researchers found that the association between phthalate exposure and dining **3**_____ was significant for all age groups, but the magnitude of association was highest for adolescents. Teens who were high consumers of foods purchased outside home had higher levels of phthalates than those who ate home-**4**_____ food. "Pregnant women, children and teens are more vulnerable to the toxic effects of hormone-disrupting chemicals, so it's important to find ways to limit their exposures," says Varshavsky. If verified by more research, these findings suggest that **5**_____ out results in higher phthalate levels.

The **6**_____ team assessed real-world exposures to multiple phthalates with a technique called cumulative phthalate exposure. This **7**_____ accounts for some phthalates being more toxic than others. In 2008 the National Academies of Sciences recommended using cumulative risk assessments to estimate the human health risk posed by phthalates; in 2017 they reported that certain phthalates are presumed to be reproductive hazards to humans.

Many products contain phthalates, including take-home boxes, gloves used in handling food, food processing equipment and other items used in the production of restaurant and fast **8**_____ meals. Previous research suggests these chemicals can leach from plastic containers or wrapping into food.

"Preparing food at home may be a win-win for consumers," says Zota. "**9**_____ cooked meals can be a good way to reduce sugar, fats and salt. This study suggests it may not have as many phthalates as a restaurant meal." At the same time, phthalate contamination of the food supply is also a public health problem that must be addressed by policymakers. Zota and Woodruff's previous research shows that policy actions, such as bans, can help reduce **10**_____ exposure to harmful phthalates.

Don't Waste Your Money on DNA-Based Diets (6)

Directions: Fill in a word that best completes each sentence.

In recent years, some companies have promoted "DNA diets" that are meant to help people lose weight by following a **1**_____ that's "personalized" to their own unique genetic makeup instead of trying a one-size-fits-all approach. However, researchers at Stanford University found that overweight adults who followed a low-fat or low-carbohydrate diet tailored to their genetic predisposition and biological makeup weren't any more successful at shedding pounds than the groups that followed the same two diets, but without the customization for these predispositions. Their findings were published in the journal JAMA.

Lead study author Christopher Gardner, of the Stanford Prevention Research Center, said that the goal of the study was to explore which factors, genetic patterns and insulin resistance, predict success for people on the two **2**_____. The researchers were trying to find out "Which diet is best for whom?" Figuring out which diet is best for **3**_____ is like "personalizing" an individual's diet.

In the study, the researchers tracked about 600 overweight adults, ages 18 to 50, who were randomly assigned to follow either a 20-gram (low) fat diet or a 20-gram (low) carbohydrate diet for one year. All the research subjects had their DNA tested before the study to see if they had one of three genes that could predict whether they might achieve better weight-loss results on a diet that was low-**4**_____ or low-carb, or whether they lacked these genes. Besides genetic testing, the participants were also given a test for "insulin resistance" to determine if their body responds properly to the hormone **5**_____, which governs how easily they absorb glucose from food. Previous research has suggested that people with greater insulin **6**_____ may have better success with a low-**7**_____ diet, because it provides a lower amount of glucose than a low-fat diet, which contains more carbohydrates.

The study found that after one year on either diet participants lost a about 10 lbs. (4.5 kilograms). Researchers did not find that being assigned to a diet that matched that individual's genetic makeup or insulin resistance could predict weight-loss success. Further, the researchers could not confirm another previous study that found that following a low-fat or low-carb diet that matches a subject's genotype results in more weight **8**_____ than subjects following diets not matched to their genetics.

Gardner said that these findings do not eliminate the possibility that there are other genotype patterns that could be useful to predict **9**_____-loss success. The Stanford researchers said they would continue to investigate what other factors might help predict an individual's weight-loss **10**_____. Characteristics such as the ability to stick to a diet, the composition of gut bacteria and psychological traits that could influence eating behavior may provide other hints about how to personalize diet recommendations, Gardner said.

Do Vitamin Supplements Improve Your Health? (7)

Directions: Fill in a word that best completes each sentence.

The body cannot create vitamins. It must get the 13 vitamins it needs to survive from outside sources. Also called "micronutrients" because they are required in minute quantities, **1**_____ can be grouped in two categories: fat-soluble vitamins A, D, E and K can accumulate in the body when consumed in excess and water-**2**_____ vitamins C and B, which are easily excreted during urination.

The best way to get vitamins is through food according to Susan Taylor Mayne, a professor at the Yale School of Public Health. For example, the vitamins found in fruit, vegetables and other foods come with thousands of other phytochemicals (plant nutrients) which may protect against cancer, cardiovascular disease, and other chronic ailments. Carotenoids in carrots and isothiocyanates in broccoli are just two examples.

The combined effect of all these vitamins and **3**_____ seems to have much greater power than one nutrient supplement taken alone, Mayne explains. For example, lycopene, the **4**_____ that gives tomatoes their red hue, has been associated with a lower risk for prostate cancer. Research suggests that taking it in supplement **5**_____ does not confer the same benefit as eating tomatoes or tomato products, such as pasta sauce and ketchup, that preserve some of the tomato's chemical integrity.

Meir Stampfer, of the Harvard School of Public Health in Boston, recommends that healthy adults take a multivitamin and extra vitamin D, if they don't get a lot of sun. Taking more than the Institute of Medicine's recommended daily allowance (RDA) of certain vitamins may lower one's risk for certain chronic diseases, he says. For example, Stampfer's research suggests that men and **6**_____ taking vitamin E supplements for years at a time have a lower risk for heart disease. "The evidence for benefit is weak," but he says that taking up to 600 IUs (international units) per day is harmless. (The RDA levels for vitamin E are 22.5 IUs.)

Mayne disagrees, pointing to a recent meta-analysis suggesting that vitamin **7**_____ supplementation increases mortality (death) of all causes. Mayne contends that there is no need to consume more than the RDA of vitamins and there is evidence that excessive intake of certain micronutrients is harmful. Stampfer acknowledges that overdosing on certain vitamins can be dangerous.

Both Mayne and Stampfer agree that more research is needed to determine the health effects of vitamin supplements—and that such **8**_____ are critical for certain people. For example, pregnant women, and women who want to get **9**_____, should be taking folic acid supplements to help prevent birth defects in their babies and people over 50 can benefit from B12 supplements because absorption of this vitamin in the digestive tract becomes less efficient with age.

Ironically, "the people who are most likely to take vitamin supplements are the **10**_____ who least need them," Mayne says. It may not be doing any good, and it could be harming them, she says.

Could Eating More Seafood Help Couples Conceive? (8)

Directions: Fill in a word that best completes each sentence.

Couples may have an easier time conceiving if both the man and the woman eat seafood more frequently, a new study suggests. In the study, published in The Journal of Clinical Endocrinology and Metabolism, the researchers found that couples who ate two or more 4-ounce servings of **1**_____ a week had sex more often, and tended to get pregnant faster, than couples who ate **2**_____ seafood. However, the study only noted an association among seafood intake, sexual activity and fertility; it didn't prove causation.

There has been little previous research on the possible links between seafood and fertility, but the few studies that have investigated the relationship have focused on the potential harms of seafood on **3**_____, such as exposure to mercury and other environmental chemicals that could have reproductive consequences, according to the study authors. These concerns may have led some women to shy away from eating fish when attempting to become pregnant, the researchers added.

In the study, the researchers looked at data collected from about 500 couples in Texas and Michigan who were trying to have a baby and were not being treated for infertility. At the beginning of the study, the researchers interviewed each partner separately, asking how often he or she had consumed seafood over the past 12 months. The participants also kept daily records of their food intake at various points during the yearlong study. After one year, 92 percent of the couples in which both partners had consumed more than **4**_____ servings of seafood a week became pregnant, compared with 79 percent of couples in which both partners had consumed two or fewer **5**_____ a week.

Oysters have a reputation as an aphrodisiac (a substance that stimulates sexual desire) but there is no scientific evidence that this is the case. It is not clear why seafood may influence a couple's sexual activity and odds of becoming **6**_____. It could be that the omega-3 fatty acids found in some fatty fish, such as salmon and sardines, could have beneficial effects on semen quality in men, as well as ovulation and embryo quality in **7**_____, the researchers suggested.

These findings are inconclusive because researchers don't yet know whether the observed benefits resulted from eating more seafood or from reducing the intake of other fatty foods and having a better overall diet, said Dr. Frederick Licciardi, a fertility expert who was not involved in the study. One major weakness of the study is that it didn't consider the fertility history of the men and women. For example, women in the study who consumed more seafood and got pregnant faster tended to be slightly older, on average, than **8**_____ who ate less fish. **9**_____ women may be more likely to have other children, which suggests that they have a proven fertility **10**_____, Licciardi noted. That could explain why it took these couples less time to become pregnant.

'Lipid Asymmetry' Plays Key Role in Activating Immune Cells (9)

Directions: Fill in a word that best completes each sentence.

A cell's membrane is its natural barrier between the inside of a cell and the outside world. It is composed of a double layer (bilayer) of lipids (such as fats, waxes, sterols, or fat-soluble vitamins). Intriguingly, it's been known for decades that the layer facing the inside of cells is made of different **1**_____ than the outside-facing **2**_____. We say that the lipid layers are asymmetrical, or unequal.

This "lipid asymmetry," or lack of symmetry (equality), is regulated by a variety of proteins. Maintenance of the asymmetry on either side of the bilayer demands a high amount of energy from the cell. It is vital to the cell's function. Dying cells, which permanently lose their lipid **3**_____, are targeted by the immune system for elimination.

Because different lipids create membranes with different physical properties, a group of McGovern Medical School researchers wondered whether different lipid compositions in the bilayer could also lead to different physical **4**_____. At the 62nd Biophysical Society Annual Meeting, in San Francisco, California, Joseph H. Lorent, a postdoctoral researcher, and Ilya Levental, an assistant professor, presented their work exploring lipid asymmetry's role in immune cell activation.

Scientists have used fluorescent probes to visualize the general **5**_____ properties of membranes. However, probes tend to stain both sides of the plasma membrane, making it impossible to independently measure the **6**_____ layers of the bilayer. Lorent and **7**_____ instead injected the probe inside of cells. Lorent explained that they used a glass syringe, "like the kind used for extracting nuclei out of cells for cloning." Lorent further explained, "This allows us to visualize specifically half of the membrane facing the [inside of the cell]." This allowed the researchers to see the changes between the inside and **8**_____ bilayer. "The differences were obvious and striking. By preventing the loss of membrane asymmetry, we inhibited the immune response," said Lorent.

The team found that adjusting the lipid asymmetry of the **9**_____ was important to the immune cells functioning. "In the long run, by knowing how lipid asymmetry is involved in cell signaling, we might be able to 'tune' certain **10**_____ responses or even cell death through the regulation of lipid asymmetry," Levental said. "This might involve treatments for allergies, inflammation or possibly even cancer."

Researchers Defy Biology: Mice Remain Slim on Burger Diet (10)

Directions: Fill in a word that best completes each sentence.

We are our own worst enemy when it comes to developing obesity. The body is naturally geared to assimilate energy from the food we eat and store it as fat until it is needed. This is the result of millions of years of evolution under the pressure of low food availability. But today, many of us have constant access to high calorie foods. Our body's ability to convert **1**_____ into fat has become problematic. Consequently, the number of overweight people is skyrocketing, leading to large health consequences for both the individual and society.

To combat this, researchers at the University of Copenhagen have managed to inhibit the body's ability to store **2**_____. The scientists have genetically deleted the enzyme NAMPT in fat tissue of mice. This has rendered the animals completely resistant to becoming overweight, even on a fatty diet.

Several previous studies have shown that the presence of large amounts of the enzyme **3**_____ in blood and in stomach fat tissue is significantly connected with humans being overweight. This study provides the first evidence that NAMPT is absolutely required to become overweight and that lack of NAMPT in fat tissue fully protects against obesity.

Researchers compared how normal mice and **4**_____ lacking NAMPT in fat tissue gained weight when given either high-fat food or a healthier, lower-fat diet. When on the healthy **5**_____, there was no difference in body weight or the amount of fat between the normal mice and the **6**_____ lacking NAMPT. However, when the mice were given high-fat food, the control mice became very obese, yet the mice lacking NAMPT gained no more weight from **7**_____-fat food than when they were on the healthier diet. In addition, the mice lacking NAMPT maintained better control of blood glucose than normal mice when eating high-fat food.

NAMPT appears to increase the metabolic functionality of almost every tissue in the body in which it has been studied. For example, there are indications that the liver and skeletal muscle may benefit from increased NAMPT activity. NAMPT is critical for fat tissue function. Unfortunately, that **8**_____ is efficiently storing fat. NAMPT in fat **9**_____ was likely once an extraordinary benefit to our ancestors. Today, however, it poses a liability.

Decreasing NAMPT may not be a viable treatment strategy in humans. There would be too great a risk for potentially harmful consequences in other tissues of the body. However, this study may pave the way for more research into how NAMPT is linked to the storage of fat from the food we eat. By learning how **10**_____ become obese, we may be able to target one of the underlying mechanisms to treat obesity and metabolic disease.

New Strains of Staple Crops Serve Up Essential Vitamins (11)

Directions: Fill in a word that best completes each sentence.

Millions of people don't have enough food to survive. Many millions more have **1**_____, but their diets lack vitamins and essential minerals, which can make them vulnerable to infections, weak bones or muscles, and problems with vision or mental health.

Researchers are trying to produce nutritive foods by developing enriched versions of staple crops, such as corn, wheat, and rice. Staple **2**_____ are those that make up the dominant parts of a population's diet. 'We need sustainable agriculture to feed the growing population with adequate nutrients, besides just enough calories,' said Dr Swati Puranik. She aims to develop calcium-rich finger millet, a hardy cereal that grows in areas of low rainfall where many other grains would fail.

Dr Puranik is researching millet germplasm variations to find those with higher calcium content. She is also checking gene markers for iron and zinc, as well as 'antinutrient' compounds which interfere with the body's absorption and use of vitamins and minerals. Dr **3**_____ and her collaborators will use breeding techniques to come up with varieties of finger **4**_____ that contain higher levels of calcium and vitamins, without using genetic engineering.

Vitamin and mineral supplements can help overcome dietary deficiencies, but Dr Puranik says, "Developing improved food crops has benefits for farmers and their families, both economic and nutritional and ultimately these calcium-rich products should have an impact in lowering rates of osteoporosis and calcium malnutrition in children or pregnant and lactating women."

In a related project, professor Paul Christou has genetically engineered maize (corn) and rice to boost vitamin A, folic acid and vitamin C, as well as many vitamins and minerals. He sees value in conventional breeding to develop fortified crop varieties, but believes genetic **5**_____ is the only current way to deliver a **6**_____ crop that meets the recommended daily amounts of vitamins and **7**_____ simultaneously.

'To my mind, in order to be successful in biofortification programs, you need to address the micronutrient deficiencies in as complete a manner as possible,' said Prof. Christou. The genetically modified maize (**8**_____) varieties have more micronutrients and an increased resistance to insect pests and parasitic weeds. The research has even opened the possibility of reducing the uptake of undesirable minerals or elements by cereal crops, such as cadmium, a heavy metal that can stunt brain development in children.

There are prospects for bringing **9**_____ modified/engineered crops to developing countries on humanitarian grounds. Genetically **10**_____ (GM) cereal varieties could have a major impact if they are accepted. But Prof. Christou recognizes that GM crops *are not* always accepted, even when they can improve nutrition for many people.

Can Eating Pasta Really Help You Lose Weight? (12)

Directions: Fill in a word that best completes each sentence.

Pasta, the often-demonized carbohydrate could "help you lose weight." A meta-analysis, published in the journal *BMJ Open*, found that eating pasta was not linked with weight gain when it was consumed as part of a "low-glycemic-index" diet. (Foods that have a low **1**_____ index release sugar slowly into the bloodstream. Pasta has a relatively low glycemic **2**_____ compared with other refined grains, such as white bread.)

The researchers analyzed data from 29 studies with a total of nearly 2,500 people who either consumed pasta as part of a **3**_____-glycemic diet or ate other carbohydrates that had a higher glycemic index. After about 12 weeks, those in the pasta groups lost, on average, about half a kilogram, or 1.1 lbs., compared with the groups that ate the other **4**_____.

The findings are not an invitation to gorge on pasta. Participants in the study's pasta groups certainly weren't doing so. Instead, they had, on average, 3.3 servings of **5**_____ each week. One serving of cooked pasta is about a half cup. Additionally, the findings apply only to pasta eaten in the context of a low-glycemic-index **6**_____. What's more, the researchers noted that the amount of weight loss seen in the study was small and that it's unclear if people would keep this **7**_____ off over the long term.

Heather Mangieri, a registered dietitian and nutrition consultant said that the results did not surprise her. "If your pasta is portioned properly and paired with a nutrient-rich vegetable and a lean protein, it can be a very healthy option," Mangieri said that a proper portion of pasta is about one-half cup to one **8**_____. She added that people need to be careful when they combine pasta with items like rich sauces. "The calorie count climbs quickly when your pasta is covered with creamy sauces and eaten with high-fat meatballs and garlic bread." When people eat pasta, they should keep in mind the glycemic index and calorie count of the other foods they add to it.

The findings highlight an important aspect of diet planning, which is that you don't necessarily need to cut out your favorite foods to maintain a healthy weight. "Good things come when you learn how to eat your favorite **9**_____ in a way that helps you maintain a healthy weight, versus depriving [yourself] and feeling as if you're on a diet," said Mangieri.

The meta-**10**_____ wasn't funded by the pasta industry; however, some of the authors previously received research grants, in-kind donations of pasta for studies, and travel support from the pasta maker Barilla.

Some Vitamins and Minerals May Carry More Risks Than Benefits (13)

Directions: Fill in a word that best completes each sentence.

Vitamins and minerals, marketed to keep you healthy, may carry more risks than **1**_____, especially as we age. "Supplements are most useful when they're used to replace dietary deficiencies," says Consumer Reports' chief medical adviser, Marvin M. Lipman. "Therefore, most of us don't need them. Such needless use can be harmful, especially if you also take prescription medications." The evidence supporting supplements is mixed and ingredients vary because of lax regulation. Several may be harmful if you're older than 50:

Folic acid (vitamin B9) may be needed by pregnant women, or women planning to get **2**_____, to prevent birth defects. Folic acid has been suggested, but not proved, to help ward off Alzheimer's disease, depression, and heart disease. A study published by the American Journal of Clinical Nutrition links excess folate (including folic **3**_____) to burning, tingling or numbness in the extremities of people with a common gene variant.

In addition, taking as little as 300 mcg daily may mask a B12 deficiency, which is relatively common in older adults, says **4**_____ Reports' medical director, Orly Avitzur. "Undiagnosed, that can lead to nerve damage, cognitive trouble and even psychiatric problems," she says. **5**_____ acid can also reduce the effectiveness of the seizure drug fosphenytoin and the cancer drug methotrexate.

Calcium supplements are often taken to strengthen your bones, which can weaken with age. People who eat little or no calcium-rich food, such as dairy products and leafy vegetables may need to take **6**_____. However, regular use of calcium **7**_____ may increase the risk of kidney stones and heart disease. A study found that people who took calcium supplements over a 10-year period were more likely than others to accumulate the artery plaque that can lead to heart attacks. Supplemental calcium can also negatively interact with some heart and thyroid medications.

Anemia, or low blood levels of iron, is more common with age. Iron supplements should be taken if diagnosed with iron-deficiency anemia. Taking too much **8**_____, however, can mask symptoms of anemia, which can be caused by internal bleeding, and lead to a missed diagnosis. Iron supplements can also inhibit absorption of certain antibiotics and blood-pressure-lowering drugs. Additionally, if you have hemochromatosis, a common genetic condition, iron can be overloaded in vital organs, potentially causing diabetes symptoms, heart problems and liver damage.

Vitamin E supplements are said to help prevent cancer, dementia and heart disease, but there's little proof. Research has linked regular use of vitamin **9**_____ to a higher risk of heart failure in certain populations. A 2011 study published in JAMA also found that taking 400 international units daily may boost the likelihood of prostate cance. **10**_____ E supplements may also make some chemotherapy drugs less effective. Consumer Reports' experts don't recommend vitamin E supplements for anyone.

Cellular Energy

Could Scientists Have Found Multicellular Animals That Live in Oxygen-Free Environment? (1)

Directions: Fill in a word that best completes each sentence.

In the 236 years since oxygen was identified as a necessity, no scientist has discovered a multicellular animal capable of living without it. Yet, researchers from the Polytechnic University of Marche in Ancona, Italy think they've discovered three new species that live in an anoxic (low oxygen) pit beneath the Mediterranean Sea.

The **1**_____, called the L'Atalante basin is off the western coast of Crete. The inner part of the basin is completely devoid of oxygen because ancient salt deposits buried beneath the seafloor have dissolved into the ocean, causing the water to become extra salty and dense.

The discovery of these three new **2**_____, if found to truly be anoxic, will drastically revise science's understanding of where animals can thrive. Prior to this discovery, the only organisms capable of life in oxygen-free environments were viruses and bacteria. Other organisms including plants, all previously discovered animals, and fungi need **3**_____. They use mitochondria, a cellular organelle, to perform the process of cellular respiration. During cellular **4**_____, sugar and oxygen are converted into water, CO_2 and, energy. The **5**_____ is used by most organisms to power their cells.

Scientists making the discovery say the three newly discovered animal species don't use mitochondria. Instead of **6**_____, the weird creatures have an organelle that resembles a hydrogenosome, a cellular component used by some microbes to produce energy with complex enzymatic reactions.

The organisms all belong to the phylum Loricifera. They measure less than 0.04 inches long. They were found almost 10,000 feet down in sediment previously assumed to contain only viruses and bacteria.

This is not the first time that scientists have discovered animals living in an anoxic environment, but all the previously discovered species needed to surface periodically for oxygen. These creatures could be the first animals ever discovered that spend their whole lives without oxygen.

The researchers published their work in 2010 in the journal *BMC Biology*. Some other researchers are not convinced that these organisms truly live without oxygen. A second team visited the Mediterranean in 2011 to examine for themselves the loriciferans and their unusual environment. Their findings, which were **7**_____ late in 2015, challenge the idea that the **8**_____ really do live without oxygen.

The scientific community is still looking for more evidence confirming or disproving the original finding. It is currently at a stalemate. Scientists need more samples for closer study. Final proof would be seeing the animals swimming around in the mud, but the small size of loriciferans and their difficult-to-reach environment makes it hard to make those sorts of observations. If loriciferans are found to be able to **9**_____ without oxygen, scientists will need to rethink current understandings of animal **10**_____.

Artificial 'Plants' Could Fuel the Future (2)

Directions: Fill in a word that best completes each sentence.

University of California researchers have combined nanoscience and biology by creating artificial plants that make gasoline and natural gas using sunlight. Peidong Yang and his team created an artificial leaf that produces methane, the primary component of natural gas. To do this, they used a combination of semiconducting nanowires and bacteria. The research is detailed in the Proceedings of the National Academy of Sciences. Yang and his colleagues previously created a similar system that yielded butanol, a component in gasoline. The researchers hope that, one day, we may be able to use those **1**_____ plant-produced fuels to heat our homes or run our cars so that we don't add greenhouse gases to the atmosphere.

The research is a major advance toward synthetic photosynthesis, a type of solar power based on the ability of **2**_____ to transform **3**_____, carbon dioxide and water into sugars. Instead of sugars, however, **4**_____ photosynthesis seeks to produce liquid fuels that can be stored for months or years and distributed through existing energy infrastructure.

In a roundtable discussion on his recent breakthroughs and the future of synthetic **5**_____, Yang said his hybrid inorganic/biological systems give researchers new tools to study photosynthesis -- and learn its secrets. "We're good at generating electrons from light efficiently, but chemical synthesis always limited our systems in the past. One purpose of this experiment was to show we could integrate bacterial catalysts with semiconductor technology. This lets us understand and optimize a truly synthetic photosynthesis system."

"Burning fossil fuels is putting carbon **6**_____ into the **7**_____ much faster than natural photosynthesis can take it out. A system that pulls every carbon that we burn out of the air and converts it into fuel is truly carbon neutral," added Thomas Moore, who also participated in the roundtable **8**_____. Moore is a professor of chemistry and biochemistry at Arizona State University.

Ultimately, researchers hope to create an entirely synthetic system that is more robust and efficient than its natural counterpart. To do that, they need model systems to study nature's best designs, especially the catalysts that convert water and **9**_____ dioxide into sugars at room temperatures.

"This is not about mimicking nature directly or literally," said Ted Sargent, another **10**_____ discussion participant, from the University of Toronto. "Instead, it is about learning nature's guidelines, its rules on how to make a compellingly efficient and selective catalyst, and then using these insights to create better-engineered solutions. Today, nature has us beat," Sargent added. "But this is also exciting, because nature proves it's possible."

Plants Use Light Trick to Make Bees See Blue (3)

Directions: Fill in a word that best completes each sentence.

A plant's reproductive goal is to swap genes with other members of their species. To do this, they need to attract pollinators so that their pollen gets carried to another **1**_____. Plant blooms are intended to draw insects like bees, but insects see best at the blue end of the spectrum. Yet blue flowers tend to be rare because the production of that color requires a lot of a plant's energy. Cambridge plant scientist Beverley Glover and her colleagues were interested in how plants were able to attract **2**_____ without producing **3**_____ flowers.

A recent article in *Nature* describes how the research team looked at several unrelated plant species and noticed that some plant petals have an iridescent effect, like a reflection from the surface of a CD, when light falls on them. "We realized that [plants] also scatter blue light off the petal surfaces to produce a blue halo around the flowers. We found that cells on the surfaces of the petals have a finely wrinkled surface. The wrinkles are the same size as the waves of blue light, which is why the blue color gets scattered off the petal surface," explained Glover.

The Cambridge team examined the diverse collection of plant species at the Cambridge University Botanic Gardens, as well as specimens from Kew Gardens in London. Many of the **4**_____ they studied also had cells in their petals capable of scattering blue **5**_____ in this way. The way the petals were scattered always resulted in the appearance of a subtle blue halo around the flower. However, the individual shapes and structures of the **6**_____ responsible were different among the different species of plants. "That told us that all these different plant species had independently evolved this trait, because they were all doing it slightly differently."

Glover and her colleagues suspect that, because blue is hard for a flower to make, plants are using this trick of light from their petals to make their flower look blue to a **7**_____ and thereby boost their pollination prospects. To investigate this, the research team made mock flowers with surfaces etched at the nanoscale in an identical pattern to the wrinkles in the **8**_____ surfaces so that the mock flowers produced the same blue halo seen around real blooms. These fake flowers were charged either with a sweet sugary treat, or a bitter **9**_____. Bees exposed to the fake flowers learned to avoid the ones with bitter nectar, or to target only the sugar-loaded examples, proving that they could clearly see the blue effect. The bees were also more time-efficient at visiting flowers with the blue halo effect.

"We've not yet proved in the wild, with real flowers, that producing this blue **10**_____ makes the flowers more attractive to bees," acknowledges Glover, "but we might be able to set up some experiments with mutant plants that either do or do not make the halo effect, so we can test that in future."

Corpse Flower (4)

Directions: Fill in a word that best completes each sentence.

One of the world's largest (10-15 feet tall) and rarest flowering structures, the corpse flower (*Amorphophallus titanium*) is a pungent (strong smelling) plant that blooms rarely and only for a short time. There is a reason for the plant's strong odor: "It all comes down to science," said Tim Pollak, of the Chicago Botanic Garden. "The smell, color and even temperature of **1**_____ flowers are meant to attract pollinators and help ensure the continuation of the species."

Said Pollak, "The insects think the flower may be food, fly inside, realize there is nothing to **2**_____, and fly off with pollen on their legs. This process ensures the ongoing pollination of the species. Once the flower has bloomed and **3**_____ is complete, the flower collapses."

The corpse flower is an inflorescence — a stalk with many flowers. A mixture of tiny male and **4**_____ flowers grow at the base of the spadix, the central phallus-like structure, which is surrounded by the spathe, a pleated skirt-like covering that is bright green on the outside and deep maroon **5**_____ when opened. If pollinated, the spadix grows into a large club-like head of orange-red seeds.

The plant's energy is stored in the corm – a swollen stem base typically weighing around 100 lbs. During the non-flowering years, a single leaf, the size of a small tree, shoots up from the corm. This **6**_____ branches out into three sections with each of these sprouting more leaflets. Each year, this shooting leaf dies and a new one grows in its place. After many **7**_____, the plant finally gathers enough energy to bloom, and once it does, it can only hold the bloom for 24 to 36 hours before it **8**_____.

Once the blooming begins, it occurs in two stages on consecutive nights: essentially a "female" stage and a "**9**_____" stage. The female flowers form a ring at the bottom of the spadix (inner tube structure), and the male flowers form a ring around the spadix just above the female flowers. During the first stage, carrion beetles drawn by the stench of death and human-like body temperatures, creep inside the vase-like structure and unknowingly deposit pollen on the receptive female flowers. During the second stage, the structure begins to collapse, the "fragrance" fades and the insects begin to head out. As they leave, the beetles rub up against the pollen in the male flowers and are now ready to carry the pollen to another nearby female flower.

The corpse **10**_____ was first discovered in Sumatra in 1878 by Italian botanist Odoardo Beccari. The plant grows in the wild only in tropical regions of Asia. The corpse flower is classified as "vulnerable" on the International Union for Conservation of Nature's (IUCN) Red List of Threatened Plants. Its main threat is habitat loss and destruction. As of now, the Sumatran rainforests are under major threat of deforestation as huge areas are logged for timber to clear space for palm plantations.

Do Fish Get Out of Breath? (5)

Directions: Fill in a word that best completes each sentence.

Like humans, fish need oxygen. And, like humans, sometimes they run short of it. When humans need to produce more ATP for energy, their breathing rate increases to deliver more **1**_____ to body cells. If the human body's **2**_____ require more oxygen then is available, the body begins using anaerobic respiration (which doesn't require oxygen) to provide the **3**_____ it needs. Unfortunately, a byproduct of anaerobic respiration in humans is lactic acid, which decreases the efficiency of muscles and can cause muscle cramps.

Fish are exposed to large oxygen fluctuations in their aquatic environment. Fish respond to hypoxia (oxygen deficiency) with varied behavioral, physiological, and cellular responses to maintain homeostasis. The biggest challenge fish face when exposed to low oxygen conditions is maintaining metabolic energy balance, as 95% of the oxygen consumed by **4**_____ is used for ATP production through the electron transport chain.

Since there are many species of fish, the way their bodies deal with energy varies. Jeffrey Richards, a biochemist at the University of British Columbia who studies hypoxia and stress, notes that there are roughly 30,000 different fish **5**_____. They each respond to exertion differently. Salmon, for example, work hard when they traverse rapids or waterfalls, but it doesn't leave them gasping. The effort is more like lifting weights than running, says Richards. The fish's muscles will tire before they run out of oxygen.

That is different for tropical fish. Salmon live in cool water, which typically has lots of dissolved oxygen; tropical fish usually live in oxygen-poor environments. Tropical fish have special adaptations to being short of breath. Some species engage in "aquatic surface respiration." They swim upward to obtain oxygen from the top-most layer of water. This **6**_____ contains more oxygen because it has been more exposed to the air than lower layers of **7**_____. Other species of fish—including sculpins and gunnels—pop out of the water for gulps of air. The lung fish has internal organs it uses as makeshift lungs. It absorbs oxygen into its blood through the walls of its' mouth, swim bladder, and stomach.

Like humans, fish use **8**_____ respiration when oxygen is unavailable. As in humans, this results in the accumulation of acidic waste products. Some fish have adapted traits that make them hypoxia-tolerant (able to withstand a low oxygen environment). One adaptation is a reduction in metabolism through lowering activity level. Other fish, such as the crucian carp, have adapted the ability to convert lactic **9**_____ into ethanol in the muscle and excrete it out of their gills. Although this process is energetically costly it is crucial to their survival in hypoxic waters.

Fish never get so short of **10**_____ that they pass out. Fish rest near the sea floor to save energy, or surface in search of oxygen. Richards says that it might seem risky to rest, "but if they're the only animals still alive at such low levels of oxygen, they won't be picked off by predators."

Engineers Create Plants That Glow (6)

Directions: Fill in a word that best completes each sentence.

MIT engineers have embedded specialized nanoparticles into the leaves of watercress plants inducing them to give off dim light for nearly four hours. The engineers believe that, with improvements, such plants will one day be bright enough to light a workspace. Their research study appears in the journal *Nano Letters*.

"The vision is to make a plant that will function as a desk lamp -- a **1**_____ that you don't have to plug in. The light is ultimately powered by the energy metabolism of the **2**_____ itself," says Michael Strano, Professor of Chemical Engineering at MIT. This technology could also be used to provide low-intensity indoor lighting, or to transform trees into self-powered streetlights, the researchers say.

Plant nanobionics, a new research area pioneered by Strano's lab, aims to give plants novel features by embedding them with different types of nanoparticles. The group's goal is to engineer plants to take over many of the functions now performed by electrical devices. Lighting, which accounts for about 20 percent of worldwide energy consumption, is a logical target. "Plants can self-repair, they have their own energy, and they are already adapted to the outdoor environment," Strano says. He thinks this is an idea whose time has come and that it's a perfect problem for plant **3**_____.

To create their glowing plants, the MIT team turned to luciferase, the enzyme that gives fireflies their glow. Luciferase **4**_____ acts on a molecule called luciferin, causing it to emit light. Another molecule called co-enzyme A helps the process along by removing a reaction byproduct that can inhibit luciferase activity. The MIT team put the luciferase, luciferin, and co-enzyme A each into a different type of nanoparticle carrier. The FDA approved nanoparticles help each of the components get to the right part of the plant. They also prevent the **5**_____ from reaching concentrations that could be toxic to the plants. To get the particles into **6**_____ leaves, the researchers first suspended the particles in a solution. Plants were immersed in the **7**_____ and then exposed to high pressure, allowing the particles to enter the **8**_____ through tiny pores called stomata.

The researchers' early efforts at the start of the project yielded plants that could glow for about 45 minutes, which they have since improved to 3.5 hours. The light generated by one 10-centimeter watercress seedling is currently about one-thousandth of the amount needed to read by, but the researchers believe they can boost the light emitted, as well as the duration of **9**_____, by further optimizing the concentration and release rates of the components.

The group's target is to perform one treatment on the plant and have it last for the plant's lifetime. The researchers have also demonstrated that they can turn the **10**_____ off by adding nanoparticles carrying a luciferase inhibitor. This could enable them to eventually create plants that shut off their light emission in response to environmental conditions such as sunlight.

Future Increase in Plant Photosynthesis Revealed by Seasonal Carbon Dioxide Cycle (7)

Directions: Fill in a word that best completes each sentence.

Doubling of the carbon dioxide concentration will cause global plant photosynthesis to increase by about one third, according to a paper published in the journal *Nature*. The study has relevance for the health of the biosphere because photosynthesis provides the primary food-source for animal life, but it also has great relevance for future climate change.

Vegetation and soil are currently slowing down global warming by absorbing about a quarter of human emissions of carbon dioxide. Something that absorbs carbon **1**_____ can be referred to as a "carbon sink". This land carbon **2**_____ is believed to be caused, in part, by increases in photosynthesis. It is widely accepted that plant **3**_____ will increase with increased **4**_____ dioxide, so long as nutrients, such as nitrogen and phosphorus, are not limiting factors to photosynthesis. That is, if nutrients such as nitrogen and **5**_____ are in ample supply, as well.

Global Earth System Models (ESMs) all predict that global photosynthesis will increase with carbon dioxide, but they differ by a factor of three in the size of this 'CO_2 fertilization'. The authors of the *Nature* study, which are based at DLR in Germany and the University of Exeter in the UK, have discovered that the size of the CO_2 **6**_____ is revealed by how the seasonal cycle in carbon dioxide concentration varies in the atmosphere.

Lead author of the study, Sabrina Wenzel of DLR explains: "the carbon dioxide concentrations measured for many decades on Hawaii and in Alaska show characteristic cycles, with lower values in the summer when strong photosynthesis causes plants to absorb CO_2, and higher-values in the winter when photosynthesis stops. The peak-to-trough amplitude [sizes of the increasing and **7**_____ cycles when shown on a graph] of the seasonal **8**_____ therefore depends on the strength of the summer photosynthesis and the length of the growing season."

The measurements made in Hawaii and Alaska show an increasing amplitude of the **9**_____ cycle. The *Nature* study shows a link between the increase in CO_2 amplitude and CO_2-fertilization. The authors call this an Emergent Constraint.

Co-author Professor Peter Cox, of the University of Exeter, summarizes the consequences of the study: "despite nutrient limitations in some regions, our study indicates that CO_2-fertilization of photosynthesis is currently playing a major role in the global land carbon sink. "This means that we should expect the **10**_____ carbon sink to decline significantly when we begin to stabilize CO_2."

Mistletoe Lacks Key Energy-Generating Complex (8)

Directions: Fill in a word that best completes each sentence.

Viscum album, commonly known as European mistletoe, common mistletoe, or simply, **1**_____ is native to Europe and western and southern Asia. European mistletoe is a hemiparasite, or semi-parasite, which is a plant which obtains some nourishment from its host but also performs photosynthesizes. *Viscum* **2**_____ parasitizes several species of trees, from which it obtains water and nutrients.

Two separate research papers published in *Current Biology* give some insight into the evolution of parasitic lifestyles on our planet. These two papers highlight the novel ways in which parasites adapt biochemically and genetically to out-compete their hosts, as well as other **3**_____. The research groups reported that European mistletoe doesn't have a key enzyme that breaks down sugar into usable energy. The teams also revealed how the parasitic plant works around that gap in its energy-generating machinery.

The enzyme the **4**_____ mistletoe is missing is called Complex I. It is so central to breaking down sugar into usable **5**_____ that it's found in the mitochondria of everything from jellyfish to people. Typically, mitochondria contain highly conserved protein complexes, including Complex **6**_____, which they use in a process called respiration to break down the sugar glucose into a molecule called ATP, which cells use as fuel. Previous research had revealed that the genes for **7**_____ I are missing from the genomes of mistletoe mitochondria, but there was a possibility those genes could have migrated to the nucleus of the cell at some point in the plant's evolution.

The two **8**_____ groups, which are based in Germany and the U.K., decided to look for the protein complex itself in the plant. Instead, they found hints that mistletoe goes about energy generation very differently from its multicellular brethren. In addition to lacking Complex I, for example, mistletoe mitochondria have low levels of some other complexes in the ATP assembly line. Cells pick up part of the slack by generating **9**_____ through reactions in the plasma of the cell such as glycolysis, a different way of converting glucose to energy.

"Maybe because mistletoe is a parasite and it gets lots of **10**_____ from its host then it doesn't need a high capacity for respiration," conjectures study coauthor Andrew Maclean of the John Innes Centre in the U.K.

Mitochondria Have Their Own DNA (9)

Directions: Fill in a word that best completes each sentence.

Mitochondria—the organelles that power our cells—have their own DNA. Scientists think that mitochondria were once independent organisms until, more than a billion years ago, they were swallowed by larger cells. Instead of being digested, they developed a mutually beneficial relationship with their hosts that eventually enabled the rise of more complex life, like today's plants and animals.

Over the years, the mitochondrial genome has shrunk. Most mitochondrial genes have jumped into the nucleus, even some **1**_____ that help the mitochondria function. Scientists wonder, if genes are mobile, why have **2**_____ retained any genes at all?

Iain Johnston, a biologist in the United Kingdom, and biologist Ben Williams of the Whitehead Institute for Biomedical Research, modeled the problem, mathematically comparing different hypotheses about the existence of mitochondrial DNA. They analyzed more than 2000 different mitochondrial genomes from animals, plants, fungi, and protists. They traced their evolutionary path, creating an algorithm that calculated the probabilities that different genes and combinations of genes would be lost at points in time.

Mitochondria make energy through a series of chemical reactions that pass electrons along a membrane. Key to this process is a series of protein complexes that are imbedded in the internal **3**_____ of the mitochondria. All the mitochondria's remaining genes help produce **4**_____ in some way. The researchers found that a mitochondrial gene was more likely to stay in the mitochondrion if it created a **5**_____ that was central to one of these protein **6**_____.

"Keeping those genes locally in the mitochondria gives the cell a way to individually control mitochondria," Johnston says. That local **7**_____ means that problems in one mitochondrion can be fixed internally instead of triggering a cell-wide response that might disrupt cell homeostasis.

John Allen, a biologist at University College London, who was not involved in the study, says, "I think that's a very fundamental feedback mechanism." In Allen's own research, he's found evidence suggesting that producing certain mitochondrial proteins where they're needed helps the cell better regulate energy production. Mitochondria, with their history as **8**_____ organisms, have their own command center to do this.

Johnston and Williams's model points out other factors that might be important as well. For instance, genes that encode water-repellent proteins might get stuck in transit if they were not made in the mitochondria. Additionally, genes that can withstand harsh conditions **9**_____ the mitochondria are more likely to stay there. Johnston thinks that the algorithm he and **10**_____ developed can analyze problems where individual traits are lost or gained. He hopes the model will help understand the pathways of disease progression.

Mitochondria: Cell Powerhouses and More (10)

Directions: Fill in a word that best completes each sentence.

Mitochondria are tiny organelles inside cells that are involved in releasing energy from food. This process is known as cellular respiration. It is for this reason that **1**_____ are often referred to as the powerhouses of the cell. Cells that need a lot of **2**_____, like muscle cells, can contain thousands of mitochondria. When the breakdown products from the digestion of food enter the cell, a series of chemical reactions occur in which those **3**_____ become incorporated into the universal energy supplier in cells known as ATP (adenosine triphosphate).

Apart from cellular **4**_____, mitochondria also play a key role in the aging process as well as in the onset of degenerative disease. During **5**_____ respiration, highly reactive molecules called free radicals are formed within mitochondria. Free radicals are potentially very damaging to cell components. If too many free radicals are released in the mitochondria, the damage can be severe, resulting in the death of the cell. To protect against free radical damage, mitochondria produce their own antioxidant enzymes.

Although **6**_____ radicals are damaging, they have an important signaling role. Scientists now believe that mitochondria operate a sensitive feedback mechanism in which some of the free **7**_____ themselves act as signals to the cell, causing the cell to calibrate and adjust cellular respiration. Therefore, completely removing free radicals would not be good for the **8**_____.

Some protection against too many free radicals may come from a plant-based diet. In the lab, antioxidants in some fruits and vegetables can neutralize free radicals. Yet, more recent research suggests that antioxidants work differently in the body than in the laboratory. It is now thought that antioxidants, such as a class of plant chemicals known as polyphenols, have a direct effect on the mitochondria. It appears that they stimulate the mitochondria to become more efficient in generating energy from food. If mitochondria are not functioning efficiently, their energy-producing capacity is reduced, more free radicals escape causing damage to the cell, and early cell death may follow. If mitochondria are more **9**_____, they generate fewer free radicals and neutralize them more quickly. The functioning of the mitochondria may be 'tuned up' by polyphenols, much like the effect induced in mitochondria by exercise.

Research over recent years is indicating that the health of mitochondria is very much lifestyle and diet dependent. Excessive consumption of sugary foods and beverages reduces mitochondrial efficiency. Lack of exercise reduces the number of mitochondria in active cells such as muscle, and they become inefficient, leaking out more free radicals into the cell. By choosing a lifestyle that includes regular exercise, daily consumption of fresh **10**_____ and vegetables, avoidance of sugary foods, control of appetite and avoiding smoking, anyone can tune up their mitochondria, which should help to promote a long and very healthy life.

Photosynthesis More Ancient Than Thought, and Most Living Things Could Do It (11)

Directions: Fill in a word that best completes each sentence.

Photosynthesis sustains life on Earth by releasing oxygen into the atmosphere and providing energy for food chains. The rise of oxygen-producing photosynthesis allowed the evolution of complex life forms like animals and land plants around 2.4 billion years ago. There is, however, a simpler form of photosynthesis that does not produce **1**_____. That simpler form of **2**_____, called anoxygenic photosynthesis, evolved first, 3.5 billion years ago, according to researchers from the Imperial College London.

An ancient bacterium, that probably no longer exists today, was the first to evolve the **3**_____ form of photosynthesis. That bacterium was an ancestor to most bacteria alive today. "The picture that is starting to emerge is that during the first half of Earth's history the majority of life forms were probably capable of photosynthesis," said study author Dr Tanai Cardona, of Imperial College London.

Scientists had previously believed that one of the groups of bacteria that still use **4**_____ photosynthesis today evolved the ability and then passed it on to other **5**_____ using horizontal gene transfer -- the process of donating an entire set of genes to unrelated organisms.

Dr Cardona created an evolutionary tree for the bacteria by analyzing the history of a protein essential for anoxygenic photosynthesis. His analysis reveals that anoxygenic photosynthesis **6**_____ before most of the groups of bacteria alive today branched off and diversified. The results are published in the journal PLOS ONE.

"Pretty much every group of photosynthetic bacteria we know of has been suggested, at some point or another, to be the first innovators of photosynthesis," said Dr Cardona. "But this means that all these groups of bacteria would have to have branched off from each other before anoxygenic photosynthesis evolved, around 3.5 billion years ago. "My analysis has instead shown that anoxygenic photosynthesis predates the diversification of bacteria into modern groups, so that they all should have been able to do it. In fact, the evolution of oxygneic photosynthesis probably led to the extinction of many groups of bacteria capable of anoxygenic photosynthesis, triggering the **7**_____ of modern groups."

To find the origin of anoxygenic photosynthesis, Dr Cardona traced the evolution of BchF, a protein that is key in the biosynthesis of bacteriochlorophyll a, the main pigment employed in anoxygenic photosynthesis. The special characteristic of this **8**_____ is that it is exclusively found in anoxygenic photosynthetic bacteria and without it bacteriochlorophyll **9**_____ cannot be made. By comparing sequences of proteins and reconstructing an evolutionary **10**_____ for BchF, he discovered that it originated before most groups of bacteria alive today.

Key Protein in Cellular Respiration Discovered (12)

Directions: Fill in a word that best completes each sentence.

We say that mitochondria function as "energy factories" for cells, but this is an over-simplification. It takes about 3000 genes to make a mitochondrion. Only about 3% of the **1**_____ necessary to make a mitochondrion are allocated for making ATP (energy). More than 95% of the genes that make mitochondria are involved with other functions tied to the specialized duties of the differentiated cell in which it resides. These other, non-ATP-related **2**_____ are involved with most of the major metabolic pathways used by a cell to build, break down, and recycle its molecular building blocks. Cells cannot make the RNA and DNA they need to grow and function without **3**_____.

Mitochondrial diseases result from failures of the mitochondria. Since mitochondria are responsible for creating the **4**_____ needed by the body to sustain life and support organ function, when they fail, less and less energy is generated within the cell. **5**_____ injury and even cell death follow. If this process is repeated throughout the body, whole organ systems begin to fail. Symptoms of mitochondrial diseases can include seizures, strokes, severe developmental delays, inability to walk, talk, see, and digest food combined with a host of other complications.

When we breathe, blood transports the oxygen we've inhaled to the mitochondria, where it is used to convert the nutrients in our food to ATP. Problems with this process, called cellular respiration, result in mitochondrial diseases, as well as diabetes, cancer, and Parkinson's. Since cellular **6**_____ is how cells get energy, organisms can't sustain life without this vital process. Even though the process of cellular respiration is basic to biological knowledge, there is much scientists have yet to understand about it. Researchers at Karolinska Institute in Sweden have discovered a new function for a protein in the process of cell respiration.

Cellular respiration depends on proteins synthesized outside the mitochondrion and imported into it, and on **7**_____ synthesized **8**_____ the mitochondrion from its own DNA. Researchers at Karolinska Institute have now shown that a specific gene (Tfb1m) in the cell's nucleus codes for a protein (TFB1M) that is essential to mitochondrial protein synthesis. If TFB1M is missing, mitochondria are unable to produce any proteins at all and cellular respiration cannot take place.

"Mice completely lacking in TFB1M die early in the fetal stage as they are unable to develop cellular respiration," says Medodi Metodiev, one of the researchers involved in the study, which is presented in *Cell Metabolism*. "Mice without **9**_____ in the heart suffer from progressive heart failure and increase mitochondrial mass, which is similar to what we find in patients with mitochondrial **10**_____." The scientists believe that the study represents a breakthrough in the understanding of how mitochondrial protein synthesis is regulated, and thus increases the chances of one day finding a treatment for mitochondrial disease, something which is currently unavailable.

The Algae Invasion (13)

Directions: Fill in a word that best completes each sentence.

About 3 billion years ago, Earth's atmosphere was mostly carbon dioxide. The microorganisms that lived on Earth survive without oxygen. It wasn't until cyanobacteria developed, that the atmosphere changed.

Cyanobacteria, also called blue-green algae or blue-green bacteria, added **1**_____ to the atmosphere as they performed photosynthesis. Organisms ate the blue-green bacteria. One of those organisms was the ancient ancestor of green algae. When the **2**_____ ate the blue-green bacteria, instead of digesting it, the ancestor absorbed it. The bacteria became a permanent resident (an invader) inside the algal ancestor. Over time, the two organisms became one.

Algae's evolutionary story is filled with invasions, like the original invasion billions of years ago. Even today, algae have invaded other species. Brent Mishler, of the University of California points out that algae are responsible for the variety of colors observed on giant clams, on coral, and even sea slugs.

Larval (immature) clams get nutrients from free-floating plankton. In adulthood, **3**_____ become sessile (immobile). In adulthood, the mantle (a tissue which extends from its body) is a habitat for symbiotic algae. By day, the clam opens its shell and extends its **4**_____ tissue so that the algae receive the **5**_____ they need to photosynthesize. The clam gets most of its nutrition from the algae. The clams and their algal partners use a collection of iridescent cells to enhance photosynthesis performed by the algae. The cells create a dazzling array of colors on the clam's mantle, including blues, greens, golds and, more rarely, white.

Most reef-building corals also contain photosynthetic **6**_____, called zooxanthellae, that live in their tissues. The corals and algae have a mutualistic relationship. The coral provides the algae with a protected environment and compounds they need for **7**_____. Zooxanthellae are brownish or green because of the photosynthetic pigment chlorophyll. They are light sensitive, increasing or decreasing in numbers based on available light. Chlorophyll can also fluoresce a deep red color under special circumstances. Corals host several types of zooxanthellae, often referred to as "Clades" which have different tolerances for temperature and light; under stressful conditions the coral may expel the **8**_____, becoming white or "bleached".

The brilliant emerald green sea slug, *Elysia chlorotica*, spends months living on sunlight just like plants because it has incorporated genes from the algae that it eats. The genes from algae allow the animal to rely on sunshine for its nutrition if something happens to their food source. *E. cholorotica* steals chloroplasts from the alga *Vaucheria litorea,* which are a part of its diet. The sea **9**_____ embeds the **10**_____ into their own digestive cells, where the organelles continue to photosynthesize for up to nine months.

Algae are invaders. Clams, corals, and sea slugs are three examples.

Does Training at High Altitudes Help Olympians Win? (14)

Directions: Fill in a word that best completes each sentence.

It's widely assumed that training on top of a mountain will give an athlete an advantage when competing closer to sea level. But it's not quite that simple. In fact, athletes are discouraged from training exclusively at high altitudes. Today's altitude training cycle comes from decades of trial and error.

1_____ training started becoming trendy with athletes after the 1968 Olympics. Some athletes thought training in Mexico City improved their performance because the elevation there is over 7,000 feet. Athletes tried training at high altitudes and then competing at lower elevations, but no significant advantage seemed to be gained from it. In the early 1990s, researchers Ben Levine and Jim Stray-Gundersen came up with a widely discussed, still slightly controversial theory: "live high, train low", meaning live at **2**_____ altitude, but train at **3**_____ altitude.

If an athlete, who lives near sea level, moves to a higher altitude to train, their body will need to compensate for the thinner air at the higher altitude. Their body will produce more erythropoietin, a hormone that tells the system to create red blood cells, which carry oxygen to muscles. Those extra **4**_____ cells running through the athlete's muscles could help increase endurance enough to change the outcome of a race.

However, at a higher altitude, the athlete who is used to sea **5**_____ conditions, will have to train slower and use less oxygen. In turn, they'll be working at a lower level than usual and lose strength. The **6**_____ blood cell count will help acclimate them to the higher altitude, but the lighter training will balance the equation. The loss of muscle strength is enough to negate any help gained from a higher red blood **7**_____ count.

In 1998, Levine explained to the New York Times that humans can never totally compensate for the altitude, and so they won't be training at their potential. That's what makes live high, **8**_____ low a vast improvement by comparison: the person living at high altitude, but training at low altitude can train as hard as they always do, and not have to take a step backward in their training so they can increase their red blood cell **9**_____.

That's still the basic premise of altitude training used today, but the finer points of the system are still in question. Robert Chapman, of Indiana University, has done work for the U.S. Olympic Committee on the live high, train low method, and he has seen improvement in elite level athletes using it. Chapman says his research has also revealed how different athletes react to this training method. Some athletes produce more **10**_____; some use a lot less oxygen at higher altitudes. There is a 1 to 1.5 percent mean improvement, but some athletes may improve 5 percent, while others may be worse off. The number of feet above sea level, the number of days spent there, and the optimum time interval to return to lower elevation for competition are all variables that require further research.

Vitamin B5 (15)

Directions: Fill in a word that best completes each sentence.

The body needs small quantities of certain organic compounds for normal growth and nutrition. These organic compounds, which cannot be synthesized by the body, are called vitamins. One type of vitamin, vitamin B, is unique because it's actually a group of vitamins. There are eight total B **1**_____, and a supplement that has all eight is called a vitamin B complex. They are: B1 (thiamine), B2 (riboflavin), B3 (niacin), B5 (pantothenic acid), B6 (pyridoxine), B7 (biotin), B9 (folic acid), and B12 (cyanocobalamin).

B vitamins turn carbohydrates into glucose, the fuel that produces energy. **2**_____ vitamins also help the body use fat and protein and are important for maintaining a healthy nervous system, eyes, skin, hair and liver. Other substances once thought to be vitamins were given numbers in the B-vitamin numbering scheme but were later found to either be manufactured by the body or not be essential for life. This is the reason why the numbers 4, 8, 10, and 11 are missing. To avoid confusion, names rather than **3**_____ are sometimes used for the B vitamins.

Vitamin B5, also called pantothenic **4**_____ and pantothenate, is vital for healthy life. Like all B **5**_____ vitamins, B5 helps the body convert food into energy. **6**_____ is found in many food sources. Some particularly good sources of pantothenic acid include mushrooms, legumes and lentils, avocados, milk, eggs, cabbage, organ meats such as liver and kidneys, white and sweet potatoes, whole-grain cereals and yeast.

Vitamin B5 provides many benefits to the body. It is found in cells as a coenzyme A (CoA), which is vital to chemical reactions. Vitamin B5 helps to: create red blood cells and stress-related and sex hormones, maintain a healthy digestive tract, process other vitamins, and synthesize cholesterol.

The Food and Nutrition Board of the Institute of Medicine created a dietary reference intake (DRI) guide for B5. The **7**_____ ranges from 1.7 milligrams per day for those under 6 years old to 5 mg/day for those 14 and older. Pregnant or breast-feeding women may need higher amounts of **8**_____ acid.

A deficiency of B5 is very uncommon. The most common side effect of pantothenic acid **9**_____ is generalized malaise (feeling discomfort or unease). Side effects can also include irritability, insomnia, vomiting, depression, stomach pains, burning feet and upper respiratory infections. Deficiency may also cause poor growth, nerve symptoms, and, uncommonly anemia. Since vitamin B5 is water soluble, excess is flushed away by the urinary tract. Very high doses of vitamin B5 (10-20 grams/day) may cause diarrhea.

Some studies have found that pantothenic acid supplements can treat or prevent disease. However, Dr. Steve Kushner, of Victory Nutrition International, said, "it is usually always better to get our vitamins from food than taking supplements when we can." Dr. Linda Girgis agrees, saying that, "when we attain these vitamins from food, they are better absorbed and metabolized than when we take **10**_____."

Classification

4 New Legless Lizards Discovered in California (1)

Directions: Fill in a word that best completes each sentence.

A pair of researchers discovered four new species of legless lizards—technically not snakes. Unlike true snakes, legless **1**_____ have eyelids and can blink. Serpents (snakes) don't have external ears, while most lizards do. Snakes tend to have longer bodies and shorter tails than their limbless reptilian cousins. Many **2**_____ lizards have tiny vestigial limbs, while snakes generally have no external appendages at all. Serpents and legless lizards look **3**_____ but are only distantly related. If you trace their evolutionary history back far enough, snakes and other legless lizard species all descended from lizards, but after they split into two different groups, the different **4**_____ each lost their legs independently. This means that while snakes are legless lizards, not all legless lizards are **5**_____.

There also tends to be ecological differences between snakes and limbless lizards. Most serpents take large prey on an infrequent basis, while lizards tend to eat large numbers of **6**_____ creatures such as insects. These differences between snakes and legless lizards are guidelines, to which there are some exceptions. Pythons and boas, for instance, have rudimentary hind limbs; Burton's legless lizard, *Lialis burtonis*, found in Australia and New Guinea, has no eyelids and eats large prey like a snake.

Functional limblessness evolved independently in several squamate reptiles (lizards, snakes and worm lizards) suggesting that the body plan offers many advantages. Snakes are the most successful of the lineages that went limbless, radiating over time into 3,000 species that have exploited nearly every available habitat. Limblessness is associated with a burrowing lifestyle. Many legless lizards spend their lives underground.

Until now, only one non-snake legless lizard **7**_____ was known to live in California. The two **8**_____ who found the new non-snake legless lizards had searched for others over the past 15 years, certain that there must be more than the just the known California legless lizard, *Anniella pulchra*. Anniella **9**_____ is a limbless, burrowing lizard often mistaken for a snake. It is about 7 inches long, with small, smooth scales typically colored silvery and yellow, although other forms exist. They live in loose, sandy soils or leaf litter.

The Los Angeles Times interviewed one of the discoverers, geologist James Parham of California State University in Fullerton, who described some of the differences between these new non-snake legless lizards, which all belong to the genus Anniella, and true snakes: "Anniella can blink at you, but snakes can't because they don't have **10**_____," said Parham. Non-snake legless lizards don't shed their skin in one piece like snakes do. Non-snake legless lizards also move differently than snakes. Snakes can coil up a lot more and are more slithery. Anniella (non-snake legless lizards) tend to be more rigid.

46 New Creatures Found in Rainforest (2)

Directions: Fill in a word that best completes each sentence.

An expedition of scientists spent three weeks in 2012 exploring rivers, mountains and rain forested areas in a south-eastern region of Suriname that has "virtually no human influence". The Conservation International team found 11 new species of fish, one new snake, six **1**_____ frogs and a host of new insects in the South American country.

Dr Trond Larsen, one of the field biologists, said they were particularly surprised by the number of frogs. "With many **2**_____ species rapidly disappearing around the globe, we were surprised and uplifted to discover so many frogs potentially new to science, including a stunningly sleek 'cocoa' tree frog," he said. It is less than 3mm long. It lives on **3**_____, using the round discs on its fingers and toes to climb.

Another of these newly discovered animals is a khaki-colored frog that has white fringes on its legs and a spur on its heel. It has been nicknamed "cowboy **4**_____." It looks like another tree frog, the "convict tree frog," but it doesn't have the convict's black and white stripes. The scientists discovered the **5**_____ frog on a small branch during a night survey in a swampy area of the Koetari River.

A new type of spiny catfish, now called the "armored **6**_____," was found and preserved as a specimen. The conservation team found a couple other types of catfish, as well.

The team also found a new species of katydid (large, long-horned grasshoppers which make a sound that resembles its name). They nicknamed it "Crayola katydid" for its bright colors. They are the only katydids known to employ chemical defenses which are effective at repelling bird and mammalian predators, according to Conservation **7**_____.

Among the other new finds were a ruby-colored lilliputian beetle (*Canthidium cf minimum*). It may be the second smallest dung beetle in south American. It was discovered in Southeastern Suriname.

In total, the researchers found 1,378 different species, and their report concluded "there are very few places left on Earth that are as pristine and untouched as this region." In fact, due to the remote nature of the area, the research **8**_____ traveled by plane, helicopter, boat and on foot, with help from 30 men from indigenous communities.

Despite the relatively pristine environment, it was not entirely free of human fingerprints: water samples showed mercury above levels safe for human consumption even though there is no upstream mining. The scientists concluded that the **9**_____ was blown in on the wind. "This demonstrates that even the most isolated and pristine parts of the world are not entirely sheltered from **10**_____ impacts. All systems are interconnected," said Larsen.

Breeding Benefits When Love Bites Wombats on The Butt (3)

Directions: Fill in a word that best completes each sentence.

Wombats are small marsupials that look like a cross between a bear, a pig, and a gopher. They are built for digging, with short legs, compact heads, short broad feet, and strong claws. Like kangaroos, koalas, and other marsupials, they have a pouch. But a wombat's **1**_____ is backwards. It opens toward the bottom, instead of towards the chest. This prevents dirt and debris from entering while burrowing.

The northern hairy-nosed wombat is an endangered species. The biggest threats to the **2**_____ are its small population size, predation by wild dogs, competition for food because of overgrazing by cattle and sheep, and disease. Australia is taking steps to protect wombats.

University of Queensland researchers in Australia have found increased pacing by female southern hairy-**3**_____ wombats is an indicator that they are fertile. In season female wombats are also more likely to bite the rumps of the males in pre-copulatory behavior at the most fertile phase of their **4**_____ cycle. This knowledge could help conservation efforts by increasing the success of captive breeding.

Associate Professor Stephen Johnston said **5**_____ were trying to better understand the southern **6**_____-nosed wombat as a breeding model for their critically endangered northern cousins. "With only about 200 northern hairy-nosed **7**_____ remaining, being able to breed these animals may one day ensure the survival of the species," he said. "There has been no captive breeding of the northern hairy-nosed wombat, and even the southern species fails to breed regularly in captivity."

Dr Johnston said the size and aggressive temperament of wombats made them difficult to work with, so behavioral indicators were a significant step forward. "We have developed a way to map the reproductive **8**_____ of the female wombat by measuring hormone levels in their urine," Dr Johnston said. "Through round-the-clock monitoring over multiple breeding cycles, we detected subtle behavioral changes associated with the fluctuations in this hormonal mapping. These behaviors could be used to identify when animals in captivity should be brought together for **9**_____, serving as cues for animal husbandry managers in zoos and wildlife facilities with southern hairy-nosed wombats."

Dr **10**_____ said that detection of the most appropriate timing for successful mating in the wombat was important not only for natural conception in wombats, but also for the next step in the research program, which involves the development of assisted reproductive techniques such as artificial insemination.

Classifying Algae (4)

Directions: Fill in a word that best completes each sentence.

Algae may be a slimy green carpet that blankets the top of ponds or neglected swimming pools; it may be long strands of seaweed in the ocean, or it may also be among the poisonous red tides of kelp that bloom near the coast and threaten marine life. Algae can be easy to see, but appearances are deceiving. Different types of **1**_____may look alike, but they're different organisms. This is the problem with algae that taxonomists have struggled with. Algae does not fit well into an organizational system.

Algae may range in **2**_____ from red to brown to yellow to green. Some types, like phytoplankton, are tiny and visible only under a microscope; other types, like giant sea kelp, can grow to more than 100 feet long. Some types of algae are unicellular; others are **3**_____.

Rick McCourt, a phycologist (scientist who studies algae) at the Academy of Natural Sciences, says that classifying algae has been a problem farther back in history than Linnaeus or even Aristotle. Algae was unique when life on Earth first began. About 3 billion years ago, Earth's atmosphere was mostly carbon dioxide. Earth's inhabitants were microorganisms such as bacteria. Cyanobacteria, also called blue-green algae or blue-**4**_____ bacteria, are likely the reason why earth's **5**_____ became habitable to current life.

"Blue-green bacteria changed the world more than any other group of organisms," says Brent Mishler, of the University of California. Blue-green **6**_____ were capable of photosynthesis. That means they started taking carbon dioxide and adding oxygen to the atmosphere. Other organisms fed on blue-green bacteria. One of these organisms, the ancestor of green algae, ate a blue-green bacterium, and, instead of digesting it, the (colorless) algal ancestor absorbed the blue-green bacterium. The bacterium became a permanent resident inside the ancestor. The two organisms each had their own set of genes, but over millions of years and many generations, the genes mixed and both sets became essential for the alga's survival. The two organisms had become a single organism. All green algae descended from that ancient **7**_____.

Other types of algae (brown, red, and diatoms) likely evolved in a similar way. "What makes all the algae groups algae is that some of the cyanobacteria went and lived inside them," Mishler says. "But they were invaded separately." Different types of algae are not related, but each type evolved from an organism that was **8**_____ by blue-green bacteria. The invaded organisms were not similar. This explains why there are many kinds of algae, and why algae are hard to **9**_____.

Some scientists advocate for a new system of classification based on evolution, rather than on current appearance and structure. It's important that scientists agree on the same names for algae in the new system. "Unless you can name something you really can't talk about it," says McCourt. "Giving a species a **10**_____ is necessary in order to study and understand what it's doing."

Mathematics Supports a New Way to Classify Viruses Based on Structure (5)

Directions: Fill in a word that best completes each sentence.

Robert Sinclair of OIST and Dennis Bamford and Janne Ravantti from the University of Helsinki have found new evidence to support a classification system for viruses based on structure. The team developed a computational tool and used it to detect similarities in the outer structures of viruses. Conventional tools had previously failed to detect **1**_____. These similarities indicate common descent.

The results, published in the *Journal of Virology*, suggest that viral **2**_____ could provide a means of categorizing viruses with their close relatives, a potentially superior approach to current **3**_____ systems. Application of this new structure-based classification **4**_____ could make it easier to identify and treat newly emerging viruses that cannot easily be classified with existing classification systems.

Viruses are difficult to classify due to their diversity, high rates of change and tendency to exchange genetic material. They have many characteristics resembling those of living things but lack the ability to reproduce without a host cell. Since they can't **5**_____ independently, they are not considered organisms. Existing classification systems are often lead to very similar viruses being categorized as different entities. These systems also can't account for the fact that viruses constantly change.

If scientists could identify a part of viral structure that can't change, that would be a basis for a better classification system. Bamford proposed classifying viruses based on similarities in their protein shell, or 'capsid'. Classification based on capsid structure would require that the amino acids in the capsid proteins be similar in related viruses. Unfortunately, conventional methods can't detect **6**_____ acid sequence similarity using viral capsids.

"The **7**_____ tools for detecting sequence similarity are very fast but they can miss things," said Sinclair. "We used a more classical approach that takes longer but is much more sensitive." The team developed the Helsinki Okinawa Sequence Similarity (HOSS) to detect amino acid sequence similarity in capsids. Using it, the researchers were able to detect similarities in sequences that conventional **8**_____ were not capable of detecting. The weak similarities suggest common **9**_____ (not convergence) as the cause for capsid similarity. This may be an aspect of viruses that is difficult to change and may provide a viable approach to classification.

Now that the researchers have shown that there are similarities between viruses that were previously undetected, further work will focus on finding more efficient methods of data extraction, beyond the HOSS. "We have also begun shifting our focus to RNA viruses, of which Zika virus and Ebola virus are examples. The genomes of RNA viruses tend to be more highly variable than DNA **10**_____, and are therefore even more challenging," says Sinclair. "But with a refined method, it [classification] could well be possible."

Differences Between Reptiles and Amphibians (6)

Directions: Fill in a word that best completes each sentence.

Herpetology which is a branch of zoology that studies reptiles and amphibians. **1**_____ includes turtles, snakes, lizards, tortoises, amphisbaenas (legless squamates), crocodiles, toads, frogs, caecilians (limbless serpentine amphibians), newts and salamanders. The name reptile refers to creeping or crawling animals. The name **2**_____ refers to dual modes of existence.

It is speculated that **3**_____ transitioned from amphibians some 50 million years ago, which perhaps explains why there are so many commonly shared characteristics. Similarities between reptiles and amphibians include that they are both: ectothermic (body temperature is regulated by external sources), chordata (possess a spinal column), can alter their skin color for camouflage thermoregulation, and have sharp eyesight. Additionally, both reptiles and amphibians use defensive traits such as: camouflage, inflation of body size, autotomization (remove tail), and mimicry to protect themselves from predation.

Among other reptilian characteristics, these are not found in amphibians: Reptiles use lungs to **4**_____. Their necks have multiple vertebra, allowing articulation. Reptiles have dry, scaly, watertight skin. Reproduction is accomplished through leathery **5**_____ laid on land or maintained inside the body. The reptile egg protects the embryo from dehydration. Reptiles have no larval stages.

Among other **6**_____ of amphibians, these are not found in reptiles: Amphibians have moist skin. They breathe via gills, lungs or through their skin. Amphibians have a single vertebra in the **7**_____ which limits head **8**_____. Reproduction is accomplished through soft eggs laid in **9**_____ or in damp media. The egg capsule is permeable to water and gases. Many amphibians undergo metamorphosis through larval stages as their lungs develop.

Both reptiles and amphibians are found in the Animalia Kingdom, within the Phylum Chordata. From there, they diverge into their own, individual Classes: Amphibia and Reptilia. It is estimated that there are more than 8,000 reptiles and 6,000 amphibians inhabiting the **10**_____. Within those populations, there is great diversity.

Two New Dog-Faced Bat Species Discovered (7)

Directions: Fill in a word that best completes each sentence.

Dog-faced bats get their name from their dog-like protruding snout. They have broad, flat skulls, small eyes and erect ears. This species exhibits sexual dimorphism, with males being slightly larger than **1**_____. Bats, in general, are estimated to live up to 20 or 30 years (extreme cases), with most averaging 4–5 **2**_____.

This species is insectivorous and forages in groups. Some may consume pollen, as well. Dog-**3**_____ bats are nocturnal and use echolocation to hunt for prey. They have been observed to roost near food sources to minimize foraging costs. **4**_____-faced bats are polygynous, with males mating with many females in a single breeding season. Successful **5**_____ results in the female producing one pup yearly.

These bats are found mostly in South American rainforests and semi-deciduous forests. For more than five decades, biologists believed that only six species of dog-faced bats existed. That number has now increased to eight with the discovery of **6**_____ new species. Both new species are described in the journal *Mammalian Biology*. The new species were discovered in the tropical forests of Central and South America.

"Identifying two mammal species new to science is extremely exciting," said study lead author Dr. Ligiane Moras, a researcher at the Universidade Federal de Minas Gerais, Brazil. "After characterizing the body shapes of 242 dog-faced bats from museum collections across the Americas and Europe, comparing their DNA, and adding in field observations including sound recordings, we consider there to be **7**_____ species in this group, two of them new to science." These newly recognized **8**_____ are the Freeman's dog-faced bat (*Cynomops freemani*) from the Canal Zone region, Panama, and the Waorani dog-faced bat (*Cynomops tonkigui*) from the eastern Andes of Ecuador and Colombia.

"The discovery of two new species of Cynomops is tremendously exciting," said Smithsonian Tropical Research Institute's Dr. Rachel Page, who was not involved in the **9**_____. "Molecular tools combined with meticulous morphological measurements are opening new doors to the diversity of this poorly understood group. This discovery begs the question: What other new species are there, right under our very noses? What new diversity is yet to be uncovered?"

"I knew we had caught something exceptional, but all the bats of the Cynomops genus look very similar in the hand," said co-author Thomas Sattler. "I had not realized we had caught a new species, now called Freeman's dog-faced bat, until Ligiane came back with the DNA results."

"We were very lucky to catch several different individuals of this species in mist nets and to record their calls," he said. "Having the call data may make it possible for us to find them again in the future and to learn more about this newly discovered **10**_____ species."

From Aristotle to Linnaeus: The History of Taxonomy (8)

Directions: Fill in a word that best completes each sentence.

Taxonomy is the study of scientific classification of living organisms. Aristotle (384-322 BC) was the first to classify animals. His groupings were based on similarities: animals with or without blood, animals that live either in water or on land. Aristotle's view of life was hierarchical. In other words, he grouped living things in order from lowest to **1**_____, with the human species being the highest. His system of classification was not based on evolution or genetic relationships. Species, in Aristotle's view, were fixed and unchanging. This view persisted for the next two thousand years.

Aristotle created the binomial system. In the binomial system, each kind of organism is defined by two names: its genus and a differing factor. Aristotle's binomial **2**_____ placed each organism in a family and then differentiated it from the other members of that family by some unique characteristic. Aristotle was not methodical in his **3**_____ system, however.

After **4**_____, there were few new discoveries in the field of biology until the 16th century AD. Explorers of that period discovered plants and animals from locations far from Europe. This excited the interest of natural philosophers (scientists). These early **5**_____ went to work naming new species and fitting them into the existing classifications. This led to new systems of classification. Many of the botanists (scientists who study plants) of this period were also physicians, who were interested in the use of **6**_____ for producing medicines.

As a youth, Carl (Carolus) Linnaeus (1707-1778) was also interested in **7**_____. At that time, there were many systems of botanical classification in use. The problem with this was many of the systems were inconsistent. The same plant could have several different scientific names, according to different classification systems.

Linnaeus became a physician as an adult but later returned to botany as his primary study. He published *Systema Naturae* in 1735. His work is notable because it provides an overall framework of classification that organized all plants and animals from the level of kingdoms all the way down to species. This system of **8**_____, although greatly modified, is the one we use today.

Linnaeus continued to improve his classification system coming closer to the system that was eventually adopted by taxonomists worldwide. Even though **9**_____ improved his own system during his lifetime, his methods have been updated since due to deeper scientific understandings. For example, whales, once considered fish, are now classified as mammals.

Today, Linnaeus is considered the father of **10**_____. He combined a hierarchical system of classification with binomial nomenclature and used it to methodically identify every species known to him. His successors have continued to revise taxonomy according to genetics. Science works when scientists stand on the shoulders of their predecessors; each reaching toward greater understanding.

How Are Animals Classified? (9)

Directions: Fill in a word that best completes each sentence.

Biologists estimate that there may be anywhere from 5 to 40 million species of organisms on earth. Those organisms make up two trillion tons of living matter, or biomass. Plants comprise over 90 percent of the biomass. Animals comprise only a small percentage of the **1**_____ (10%), but they account for most species.

In accordance Linnaeus' classification method, scientists group animals and plants based on shared physical characteristics. They place them in a hierarchy of groupings. In the kingdom animalia there are groups of phyla; in a phylum (singular for phyla) there are groups of classes; in a class there are groups of orders; in an **2**_____ there are groups of families; in a **3**_____ there are groups of genera; and in a genus (singular of genera) there are groups of species. Animals are named by their genus, capitalized, and species, uncapitalized.

The most abundant and diverse animal communities occupy earth's most biologically productive regions, such as the tropical rainforests. Conversely, the least abundant and **4**_____ animal communities live in the least biologically productive regions, such as deserts. Animals living in either a rainforest, desert, or other biome must be uniquely adapted to survival to their own particular habitat.

Taxonomists divide the animal kingdom into phyla. Simplistically, we can think of animals as either invertebrates (animals without backbones) or vertebrates (animals with **5**_____). Invertebrates are part of the phylum arthropoda (arthropods). Two of the most commonly known classes in this phylum are arachnids (spiders) and insects. The five most well-known classes of **6**_____ are mammals, birds, fish, reptiles, amphibians. They are all part of the phylum chordata.

Mammals are a class of vertebratess that includes people, dogs, cats, horses, platypuses, kangaroos, dolphins and whales. They have body hair and drink milk as infants. Birds are the class aves, which has feathers and are born from hard-shelled eggs. **7**_____ are the only vertebrates that have feathers. Fish are vertebrates that live in water and have gills, scales and fins on their body. Fish have many variations including the ability of some to crawl onto land and hop! Reptiles are a class of vertebratel with scaly skin. They are cold blooded and are born on land. Snakes, lizards, crocodiles, alligators and turtles all belong to the **8**_____ class. Amphibians are born in the water. When they are born, they breath with gills like a fish. But when they grow up, they develop lungs and can live on **9**_____.

Classes of animals are divided into orders. For example, there are eight orders of mammals, including primates, insects, and carnivores. **10**_____ (more correctly carnivora) include sixteen families. Among them are cats, dogs, and bears. Even more divisions of genera and species exist. For example, the family of cats (Felidae) includes the genera lepardus and lynx, among others. At the most specific level of species there are Iberian and Siberian Lynx, as well as the ocelot and margay— types of leopards.

How Are Plants Classified? (10)

Directions: Fill in a word that best completes each sentence.

Carl Linnaeus (1707 to 1778) laid the foundation for a system that is used for classifying organisms based on shared physical characteristics. His published work, *Systema Naturae*, is a milestone document in biology. His basic system, continually expanded and modified, is the one most commonly used today.

Five major categories, or "kingdoms," of living organisms have been established: Plantae (plants), Animalia (animals), Fungi (i.e., toadstools and mushrooms), Monera (bacteria and blue-green algae), and Protista (i.e., protozoa). Within each **1**_____, scientists have established six basic hierarchical groupings: Phyla, Orders, Families, Genera, and Species. At each **2**_____ level, from phyla to genera, scientists group organisms with increasingly closely shared characteristics. At the final level, species, organisms have very similar **3**_____.

Taxonomists using the traditional Linnaean method, separate the plantae kingdom into four major groups including: mosses and liverworts, which have no proper root systems; ferns, which have proper roots and produce spores (specialized reproductive cells); coniferous trees, which have **4**_____ systems and needle-shaped leaves and cones; and flowering plants, which have root **5**_____ and flowers that produce seeds. The dominant division in many biomes, with a quarter of a million species, is that of the flowering plants.

Depending on the method they follow, taxonomists may divide the flowering plants – the most recently evolved of the botanical divisions – into two broad groups, or classes. This first class, the "monocot," produces a single first, or "embryonic," leaf from its seed. Typically, its leaves have parallel veins, and its stems have vascular bundles (the water-conducting vessels) that occur in a random pattern in cross section. It produces flowers with petals which occur in multiples of three. The second **6**_____, the "dicot", produces two embryonic leaves from its **7**_____. It produces flowers with petals that occur in multiples of four or five. Its leaves have network veins. Its stems have vascular **8**_____ that occur in a concentric ring pattern in cross **9**_____. Taxonomists are continually updating the classifications for flowering plants, especially the dicots, as new technologies and knowledge emerge.

Taxonomists have struggled to divide the classes into orders that contain logically related groups of flowering plants. By one method, they have established super-orders, which comprise family groups thought to have evolved, along different pathways, from common ancestors. Further, they have divided the super-orders into various orders. They have, however, disagreed over the groupings and divisions. Taxonomists have had somewhat more success in classifying plants as families, genera and species.

Although plants account for 90 percent of the earth's two trillion tons of biomass, they represent only 20 percent of earth's classified and named organisms. In contrast, animals make up a small fraction of the biomass, yet they represent more than 60 percent of the classified and named **10**_____.

Nature's Misfits: Reclassifying Protists Helps Answer How Many Species Remain Undiscovered (11)

Directions: Fill in a word that best completes each sentence.

Since the Victorian era of the 1800s, categorizing the natural world has challenged scientists. No group has presented a **1**_____ as tricky as the protists, the tiny, complex life forms that are neither plants nor animals. A new reclassification of eukaryotic **2**_____ forms draws together the latest research to clarify the current state of protist diversity and categorization, as well as the many species that remain to be discovered.

"Protists include species traditionally referred to as protozoa and algae, some fungal-like organisms, and many other life **3**_____ that do not fit into the old worldview that divided species between plants and **4**_____," said Professor Sina Adl, from the University of Saskatchewan. "By the 1960s it had become clear that these species could no longer fit within such a narrow system, yet the first community-wide attempt to rationally categorize all the **5**_____ in the natural evolutionary groups was only made in 2005."

The 2005 reclassification, led by Professor Adl and published by the *Journal of Eukaryotic Microbiology*, gave scientists a structure for understanding these species; however, it was limited to the technology available at the time and recent advances have prompted the need for a **6**_____. "With environmental genomics we are experiencing a renaissance of new protist discoveries," said Adl. "These new **7**_____ allow us to better appreciate how little we know about the biodiversity around us and how they contribute to maintaining the planet's chemical balance."

The most significant changes are the introduction and recognition of six new super groups, larger than traditional biological kingdoms. This reflects a greater understanding of the most ancient relationships between protists, their shared ancestry and their connections to animals and **8**_____.

This includes recognition of the Amorphea, a super **9**_____ that links animals, fungi and their protist relatives, including the marine choanoflagellates, to a diverse group of protists largely dominated by various amoeboid cells. This includes macroscopic slime molds, shell-dwelling amoebae, small flagellated amoebae and large voracious amoeboid predators of bacteria, algae and even small crustaceans.

"This new classification, that better reflects how species are related, improves our ability to predict the number of species that remain to be **10**_____," concluded Professor Adl. "There is a huge unknown diversity in the deep sea, but probably even more in the soil we walk on."

Biologists May Have Found a New Fungal Group (12)

Directions: Fill in a word that best completes each sentence.

A paper published in *Nature* suggests that biologists in the UK have discovered an entirely new and unique branch in the tree of life. A group of mysterious microscopic organisms related to fungus are so different that they make up their own kind of fungal group. There are so many of these distinctly different kinds of organisms living in so many diverse places, that the biodiversity among this new **1**_____ might be as vast as the entire known fungal kingdom. In fact, they might not actually be fungi at all.

The scientists who have discovered this new clade (a branch on the tree of **2**_____ that consists of an organism and all its descendants) have named it cryptomycota, which means "hidden from the kingdom Fungi". Indeed, the cryptomycota have remained hidden from sight. Though, they were hiding in plain **3**_____ because they are everywhere, living in different environments, including freshwater lakes, sediments, and pond water. Crypotmycota were first detected as DNA sequences retrieved from a fresh **4**_____ laboratory enclosure. Phylogenetic analysis of those sequences determined that they were unique to any other known **5**_____.

Biologists estimate that they've only categorized and cataloged about 10 percent of all fungi in the world. **6**_____ is so biologically different than other known **7**_____ groups, scientists have yet to characterize its life cycle precisely. Scientists will need to investigate how cryptomycota evolved to survive in so many diverse **8**_____.

Cryptomycota, also referred to as Rozellida, or Rozellomycota differ from classical fungi in that they lack chitinous cell walls at any trophic stage in their lifecycle. Without the chitin the "these fungi could be phagotrophic parasites that feed by attaching to or living inside other cells," says Tim James, a fungal geneticist from the University of Michigan in Ann Arbor. According to James, the lack of a **9**_____ cell wall could let the cryptomycota feed by phagotrophy – engulfing **10**_____ and digesting it internally, as opposed to osmotrophy – taking in nutrients from outside the cell, which is what most known fungi do.

Scientists will need to determine if cryptomycota are fungi, or something completely different. To do this, they will need a lot of time to research the species. There will likely be much argument in the global biological community about this new discovery.

What do Piranhas and Goldfish Have in Common? (13)

Directions: Fill in a word that best completes each sentence.

The common goldfish, tiny minnows, the tetra, the Electric Eel, the enormous Mekong Giant Catfish, the piranha, and the zebrafish all have something in common: they are members of Ostariophysi, a group of over 10,000 species of bony fishes. They all have special bone structures that transmit sound waves from their swim bladder to their inner ear that give them hearing abilities better than humans.

Ichthyologists, scientists who study **1**_____, disagree about how this diverse group of over 10,000 species of **2**_____ can be so closely related. The scientists don't understand the evolutionary relationships among diverse fish like the electric eel and knifefish or the relationships between carps and minnows.

In a paper published in *Systematic Biology,* researchers from universities and museums across the U.S. and Mexico used highly conserved regions of genetic material, called ultraconserved elements (UCEs), to gather together a data-rich family tree of fish. The study authors, including Prosanta Chakrabarty and Brant Faircloth, wanted to uncover how this group of fish, which generally don't look anything alike, could be so **3**_____ and yet closely related.

The researchers used hundreds of UCEs (genetic elements) at a time to reconstruct a Tree of Life for the **4**_____ fish. The **5**_____ of Life for these fishes had previously been based on morphology (characteristics such as skeletal structures and features of the nervous system), or just a few genes.

UCEs are regions of the genome (DNA) which stay relatively the same over long periods of time. This makes the **6**_____ easy to find. Prosanta Chakrabarty, the study's first author and Curator of Fishes at the LSU Museum of Natural Science said, "We were starting to use UCEs to better understand deep relationships in more difficult-to-handle parts of the Tree of **7**_____."

Their research found that, even though ray-finned fishes including tetras and piranhas have similar structures (morphologies) they are not each other's closest relatives. Tetras are more closely related to catfishes. Piranhas were found to have more than a single origin.

Understanding how species on the Ostariophysi tree are related to one another can help researchers figure out how this very large group has been so successful. Using knowledge of the **8**_____ within this group of fishes that evolved on an ancient super continent may also help **9**_____ understand how a single landmass broke apart. Species **10**_____ can tell us a lot about ancient geography and other environmental conditions.

Platypus: The Unique Mammal Who Might Be Able to Fight Bacteria (14)

Directions: Fill in a word that best completes each sentence.

Ornithorhynchus anatinus, is the Latin name for the platypus. **1**_____ names were created for the entire living world by Carl Linnaeus, a Swedish botanist and zoologist, in the 1700's. The classification system identified kingdoms, classes, orders, genera, and species.

The Linnaean system wasn't perfect. The **2**_____ was often based on flawed understandings of animals. Many animals were categorized based on incorrect 18th century knowledge. For example, the Linnaean differentiation between genera and species had to do with physical features, such as fur, appendages, and egg laying. The physical feature of wings caused scientists to first group bats as closely related to birds, however, scientists now have identified bats as mammals, far more closely related to humans than to **3**_____.

No animal seems to defy classification more than the platypus. As a mammal it provides its young with milk, it has fur, and it has three specific ear bones. Platypuses are mammals by our modern definition, but they weren't by Linnaeus'. The **4**_____ definition of mammal included viviparity. A viviparous animal is one that gives birth to live young. Platypuses lay eggs. To account for platypuses and echidna (**5**_____ laying mammals), scientists invented the branch of mammals called the monotremes.

The fact that platypuses are **6**_____ (egg laying) is not the only unique feature of these mammals. Platypuses give milk to their offspring, but they don't have nipples, like most **7**_____. Instead, they "sweat" out their **8**_____ from pores along their stomachs. The platypus has a hard snout that looks like a duck's bill. Platypuses have additional bones in their shoulders not found in any other mammals and their legs are mounted from the sides of their bodies like reptiles. As a result, they swim differently from mammals. Tiny spurs on the feet of platypuses contain venom, which they use for intimidating other platypuses. Platypus eyes are unlike any other four-legged creatures; they more closely resemble those of a hagfish or lamprey. Female platypuses have only one functioning ovary, like birds. The platypus' body temperature is about 90°F, 8° less than most mammals.

In addition to all these unique qualities, an antibacterial protein, which is found in platypus milk, can kill bacteria and other microbes. This protein has a highly unusual structure that has not yet been found elsewhere in nature. The Australian researchers who found the **9**_____ called it 'Shirley Temple,' because it has ringlets. They hypothesized that, due to its unique structure, the protein might also function differently from our current antibiotics, possibly making it kill **10**_____ that other drugs can't.

This research on the platypus is a living example of how hard it is to categorize an animal based on physical attributes. Scientists still have much to learn from this creature. The platypus makes you question everything you thought you knew about what made a mammal a mammal, or a bird a bird.

A Few Bad Scientists Are Threatening to Topple Taxonomy (15)

Directions: Fill in a word that best completes each sentence.

The African spitting cobra spits toxins into its victims' eyes and can deliver a bite which can lead to death. According to the International Commission of Zoological Nomenclature (ICZN), the reptile belongs to the genus *Spracklandus*. However, most taxonomists use the unofficial **1**_____: Afronaja. This can be a problem "if you walk in [to the hospital] and say the snake that bit you is called *Spracklandus*, you might not get the right antivenin," says Scott Thomson, of the University of São Paulo.

Raymond Hoser, the researcher who gave *Spracklandus* its official name, is one of the figures in a debate within the world of taxonomy. Hoser has named over 800 taxa of snakes and lizards. Prominent taxonomists and herpetologists (scientists who study amphibians and reptiles) say the number of species named by Hoser is misleading. According to them, **2**_____ has committed taxonomic vandalism.

Linnaean taxonomy is a biological classification system that has allowed scientists worldwide to study organisms without confusion for 300 years. Taxonomy does, however, have some flaws, which are being exploited by some people in the field. "Taxonomic vandals" are scientists who name many new **3**_____ without presenting sufficient evidence for their finds. They are viewed as glory-seeking scientists who use others' original research to justify their so-called "discoveries."

Normally, after a scientist has evidence of a previously unidentified organism, obtained a specimen of it, written a paper describing the discovery and given it a taxonomic name, they submit the **4**_____ for publication in a peer-reviewed journal. Publication in a peer-**5**_____ journal can take months or years.

The ICZN requires scientists to publish discoveries of new animal taxons but does not require **6**_____-reviewed publication. That leaves room for un-ethical scientists to self-publish. Doug Yanega, a Commissioner at the **7**_____, said, "No other field of science, other than taxonomy, is subject to allowing people to self-**8**_____."

The scientific community often rejects the names that vandals ascribe. Unfortunately, this results in "parallel nomenclature": when a single taxon is known by more than one name. Confusion created by parallel **9**_____ can impact how much conservation funding an endangered species will receive, as well as make it more difficult to acquire an export permit for research, taxonomists say.

One proposed solution is for the International Union of Biological Sciences (IUBS) to form a commission that would establish hardline rules for delineating new species and take charge in reviewing taxonomic papers for compliance. That measure is unlikely to be taken, however, because any change requires consensus across the taxonomic community. Consensus is unlikely because taxonomic vandalism is not a problem in most branches of taxonomy. The area of herpetology is where many vandals operate. Herpetology is home to thousands of undescribed species. **10**_____ are still looking for a solution to this scientific crime.

Ecology

Beavers Could Help Curb Soil Erosion, Clean Up Polluted Rivers (1)

Directions: Fill in a word that best completes each sentence.

A new study Led by University of Exeter's Professor Richard Brazier has demonstrated the impact a pair of Eurasian beavers have had on reducing the flow of tons of soil and nutrients from nearby fields into a local river system. The study surveyed sediment depth, extent and carbon/nitrogen content in a sequence of beaver pond and dam structures on a controlled 1.8 ha site in South West England.

"The animals built 13 dams, slowing the flow of water and creating a series of deep ponds along the course of what was once a small stream," the scientists explained. "We measured the amount of sediment suspended, phosphorus and nitrogen in water running into the site and then compared this to water as it ran out of the **1**_____ having passed through the ponds and dams. We also measured the sediment, **2**_____ and nitrogen trapped by the dams in each of the ponds."

The results showed the **3**_____ beaver dams had trapped 101.53 tons of sediment, 70% of which was soil, which had eroded from 'intensively managed grassland' fields upstream. Further investigation revealed that this **4**_____ contained 15.9 tons of nitrogen and 0.91 tons of phosphorus, which are known to create problems for wildlife in rivers and streams and which also need to be removed to meet drinking water quality standards.

"It is of serious concern that we observe such high rates of soil loss from agricultural land," Professor Brazier said. "However, we are heartened to discover that beaver **5**_____ can go a long way to mitigate this soil loss and also trap pollutants which lead to the degradation of our water bodies."

"Were beaver dams to be commonplace in the landscape we would no doubt see these effects delivering multiple benefits across whole ecosystems, as they do elsewhere around the world," said Brazier. The benefits of **6**_____ dams have been noted in the western United States because they have beneficial effects that can't easily be replicated in other ways. They raise the water table alongside a stream, aiding the growth of trees and plants that stabilize the banks and prevent erosion. They improve fish and wildlife habitat and promote new, rich soil.

Beavers are ecosystem engineers. When a family moves into new territory, the rodents drop a large tree across a **7**_____ to begin a new dam, which creates a pond for their lodge. They cover it with sticks, mud and stones. As the water level rises behind the dam, it submerges the entrance to their lodge and protects the beavers from predators. This pooling of **8**_____ leads to a cascade of ecological changes. The **9**_____ nourishes young willows and aspens and provides a haven for fish that like slow-flowing water. The growth of grass and shrubs alongside the pond improves habitat for songbirds, deer and elk.

Brazier's study shows what **10**_____ beavers can do in a controlled environment. Beavers are important to ecosystems overall. This study's findings were published in *Earth Surface Processes and Landforms*.

Old Maps Highlight New Understanding of Coral Reef Loss (2)

Directions: Fill in a word that best completes each sentence.

Known as the "forests of the ocean", coral reefs represent an entire underwater ecosystem, teeming with life; but they are an **1**_____ which is under threat. Climate change currently threatens mass destruction of our planet's coral **2**_____.

Researchers from The University of Queensland and Colby College in the United States were interested in finding out how much coral has been lost from the earliest historical records. The team examined coral loss over a period of 240 years by comparing 18th Century nautical charts of the Florida Keys area with satellite images and other contemporary data. Their findings have been published in *Sciences Advances*.

"We know a lot about **3**_____ reef change over that last several decades, but we really don't know as much about them over the scale of centuries," explains Colby College study author Loren McClenachan. The British **4**_____ charts obtained from the UK Hydrographic Office offered a window into the past, to a time when people first began to negatively affect these ecosystems.

According to McClenachan, when the nautical **5**_____ were compared with modern satellite data, the researchers made the shocking discovery that over half of the coral mapped in the 1770s does not exist anymore. It died off before modern-day ecologists could even begin to study it. It is difficult to pinpoint exactly when the coral began to disappear, but early in the 20th century there was a lot of **6**_____ activity in the area that had the potential to harm the **7**_____, including draining the freshwater Everglades system. Earth has lost a lot more coral than scientists ever realized it had.

In causing corals to **8**_____ off, **9**_____ have lost not only beautiful environments, but food security from fish extinctions and protection from storms because coral reefs absorb the energy from big waves. Clearly, coral reefs are important for many reasons.

Attempts to preserve our remaining coral reefs are underway. Florida Keys residents are particularly active in conservation, engaging in coral out-planting, where coral is carefully grown in special nurseries before being taken out and attached to existing reefs. Other methods include protecting fish species that indirectly allow the coral to come back. "But the biggest thing we need to do is get a handle on **10**_____ change to prevent the waters from warming," says McClenachan.

What is the Environmental Impact of Your Sandwich? (3)

Directions: Fill in a word that best completes each sentence.

Researchers at The University of Manchester have carried out the first ever study looking at the "carbon footprint" (amount of CO_2 emitted into the atmosphere) of sandwiches. The researchers examined 40 different home-made and pre-packaged varieties. They found the highest carbon footprints for the sandwiches with pork (bacon, ham or sausages) cheese or prawns as ingredients. The ready-made 'all-day breakfast' sandwich (which includes egg, bacon and sausage) had the highest carbon **1**_____. The researchers estimate that this type of sandwich generates 1441 grams of carbon dioxide equivalent. This is equivalent to CO_2 emissions from driving a car for 12 miles.

A home-made ham and cheese sandwich was lowest in carbon emissions. The study found that making your own **2**_____ at home could half carbon emissions compared to ready-made equivalents.

According to the British Sandwich Association (BSA) more than 11.5 billion sandwiches are consumed each year in the UK alone. Around half of those are made at **3**_____ and the other half are bought over the counter in shops, supermarkets and service stations around the country.

Professor Adisa Azapagic, from the School of Chemical Engineering and Analytical Sciences, said: "Sandwiches are a staple of the British diet, so it is important to understand the contribution from this sector to the emissions of greenhouse gases. Consuming 11.5 billion sandwiches annually in the UK generates, on average, 9.5 million tons of CO_2, equivalent to the annual use of 8.6 million cars."

The results show the largest contributor to a sandwich's **4**_____ footprint is the agricultural production and processing of their ingredients. Keeping sandwiches chilled in supermarkets and shops, as well as packaging and transporting them, contributes to the amount of CO_2 **5**_____ put into the **6**_____. The study concludes that the carbon footprint of sandwiches could be reduced by as much as 50 per cent if a combination of changes were made to the recipes, packaging and waste disposal. The researchers also suggest extending sell-by and use-by dates to reduce sandwich waste.

Professor Azapagic, who also heads up the Sustainable Industrial Systems research group, added: "We need to change the labeling of food to increase the use-by date as these are usually quite conservative. Commercial sandwiches undergo rigorous shelf-life testing and are normally safe for consumption beyond the use-by **7**_____ stated on the label." The BSA also estimates that extending the **8**_____-life of sandwiches by relaxing such dates would help save at least 2000 tons of sandwich **9**_____ annually.

The study also recommends reducing or omitting certain ingredients that have a higher carbon footprint. Reducing **10**_____, such as cheese and meat, would also reduce the number of calories eaten, contributing towards healthier lifestyles.

The Terrifying Way Fire Ants Take Advantage of Hurricane Floods (4)

Directions: Fill in a word that best completes each sentence.

Texans escaping Hurricane Harvey's floodwaters must watch out for floating balls of fire ants. The red imported **1**_____ ant evolved in the South American floodplains, but has since moved into the southern U.S. These ants adapted to float, so they take advantage of the habitat created by floods. Even after the waters recede, fire ants are the first ones there to grab all the available resources.

When rains first submerge a colony, the ants join using their jaws and sticky pads on their legs, according to a study in the Proceedings of National Academy of Sciences. Clumps form in just about a minute and a half. Each bunch can contain anywhere from thousands to millions of **2**_____. The critters at the bottom regularly switch out with the ones on the top, so (almost) none of the insects stay under long enough to **3**_____. The most valuable members of the group—like the queen and her newest babies—are pushed to the top and center for safe keeping, while a few unlucky individuals get permanently pressed to the edge of the ball by their neighbors. If one part of the newly mobile colony is disturbed, other ants will move to fix the gap. These living rafts can stay together for days, or as long as it takes to reach a tree or dry land.

Normal ants bite and then spray acid on the new wound, but fire ants are much **4**_____. They bite, hold on, and inject a venom containing 46 different proteins, including poisons that sometimes affect the nervous system. They also have a more brutal attack pattern than many social insects. If you knock over a beehive, not all the **5**_____ will come after you—most colonies have a few dedicated warriors to protect the **6**_____. When fire ants are disturbed, however, they all attack. About one in every hundred people will have a full-body response to the **7**_____, such as an allergic reaction or even hallucinations.

The roving balls of ants can be stopped by suds. Soapy water prevents the ants from trapping bubbles of air around their bodies. Therefore, the ants sink. Ant swarms can be warded off by spraying them with detergent. Still, the best way to fight these floating menaces is to learn to identify and avoid them.

Fire ants may even cause a decline in native species, though it's tough to tell if fire ants or humans are to blame. This is because **8**_____ have replaced natural landscapes with non-native plants and materials, which also cause a decline in native **9**_____. In so doing, humans have unknowingly created perfect conditions for **10**_____ ants. Even watering your lawn creates a "fire ant welcome mat".

Fire ants represent the worst of the worst in terms of what can happen if you allow organisms to move into a location where they haven't evolved.

If People Were Cockroaches, Adapting to Climate Change Would Be Easy (5)

Directions: Fill in a word that best completes each sentence.

Depending on the level of radiation released, it is possible that cockroaches could survive a nuclear holocaust. They were reported to have survived the blasts at Hiroshima's and Nagasaki's grounds zero. What is certain is that they can survive substantially higher doses of **1**_____ than *Homo sapiens*. What's more, if both species face extinction, it's a pretty sure bet that we'll disappear before the **2**_____ will, regardless of whether the cause is radiation, climate change, or some other, yet unimagined threat to our fragile existence.

Cockroaches have found a place in virtually every climate and ecological system on the face of the planet, from tropical jungles to the ice and snow of northern Canada. We have no idea what goes on in the 99 percent of the world's **3**_____ populations that have nothing at all to do with people.

The effects of **4**_____ change will have to be much more drastic before they begin to seriously alter the world's roaches. This is because the cockroach is one of evolution's grand success stories. It is designed for survival. It can and will eat almost anything. What it can metabolize ranges from book-binding glue to human feces, dead skin and toenails, to anything rotten. Those that live far from humans can **5**_____ whatever their world provides, including rotting trees to dead animals.

Additionally, cockroach species seem to mutate readily to protect themselves. A lot of time and money is spent trying to devise ever better ways and weapons to exterminate the few species that live in our homes. Each time a substance is developed that does a better job of **6**_____ roaches, it does not take many generations of affected insects before the local roach population develops resistance. Cockroaches have evolved a **7**_____ to many classes of insecticides.

With few cultural exceptions, cockroaches have mostly been viewed negatively. Yet, they help to clean up our garbage and our waste products. Their reputation is dirty because they can carry diseases to humans, but they also **8**_____ up the forest and the kitchen. Further, they have served generations of biology students in university labs since they have a simple, easily observable nervous system.

As **9**_____ change displaces cockroaches, more of them may discover the human homes, and if so, there is no doubt that with their superior adaptability they will remain. This is just one more of the consequences the **10**_____ in climate will produce.

Inside Australia's War on Invasive Species (6)

Directions: Fill in a word that best completes each sentence.

The Australian government unleashed a strain of a hemorrhagic disease virus into the wild earlier this year, hoping to curb the growth of the continent's rabbit population. This move might sound cruel, but the government estimates that the **1**_____—brought by British colonizers in the late 18th century—chew through about $115 million in crops every year. And the rabbits are not the only problem. For more than a century Australians have battled against the waves of non-native species that have overpopulated the area with many different measures—including introducing nonnative predators—with limited success.

Australia is not the only country with invasive (non-native) creatures. But because it is an isolated continent, most of its wildlife is restricted to its borders—and its top predators are long extinct. This gives alien (non-native, invasive) **2**_____ a greater opportunity to live and multiply. "In other places, you'll see a much bigger predator community," says Euan Ritchie, one of the directors of the Ecological Society of Australia. But the Tasmanian tiger, the marsupial lion and Megalania (a 1,300-pound lizard) are gone. The only top **3**_____ left, the Australian wild dog, or dingo, has a declining population due to **4**_____ because the **5**_____ hunts sheep which humans raise.

Along with **6**_____, Australia is trying to fight off red foxes (imported for hunting), feral (wild) cats (once kept as pets), carp (brought in for fish farms) and even camels (used for traveling in the desert). Wildlife officials have attempted to fight these invaders by releasing viruses, spreading poisons, building thousands of miles of fences, and sometimes hunting from helicopters. In one famous case, the attempted solution became its own problem: the cane toad was introduced in 1935 to prey on beetles that ate sugarcane. But the **7**_____ could not climb **8**_____ plants to reach the beetles and are now a thriving invasive **9**_____ themselves.

Despite scientists' disagreement, the government plans to introduce another **10**_____ later this year to try reducing the out-of-control carp population.

For Lemurs, Size of Forest Fragments May Be More Important Than Degree of Isolation (7)

Directions: Fill in a word that best completes each sentence.

Lemurs are found almost exclusively in Madagascar, an island nation off the southeast coast of Africa. Nearly all lemur species are at risk of extinction. The reason for this is primarily due to habitat loss and fragmentation. Fragmentation is the division of **1**_____ into smaller parcels, usually because of human activity.

To better understand how forest loss and fragmentation affects **2**_____, ecologists Travis Steffens and Shawn Lehman observed six lemur species living within 42 fragments of the dry deciduous forest in Ankarafantsika National Park, Madagascar, between June and November 2011. The researchers used mathematical models to examine whether the lemurs formed metapopulations, spatially separated populations within a species, in a fragmented landscape and under different forest fragmentation conditions. Their study was published in May 2018 in the open-access journal PLOS ONE.

In their simulations, the researchers found that three of the lemur **3**_____ formed metapopulations in forest fragments. The number of individual lemurs within each metapopulation was affected by both forest fragment size and isolation. However, **4**_____ size appeared to have a bigger effect than **5**_____, with larger forest patches being associated with increased lemur occurrence.

Madagascar has a high degree of habitat transformation typically caused by humans. It is one of the best places to understand how habitat loss and fragmentation impact primates. **6**_____ are at risk for extinction mainly due to habitat **7**_____ and habitat fragmentation. Arguably, half of lemur habitat has been lost since the 1950s. However, lemurs are also at risk of extinction from hunting and capture for the pet trade.

In a continuous forest, lemurs would more readily interact with members of the same **8**_____ in different groups. However, the lemurs forming metapopulations in this study were likely more limited in their ability to move and disperse between fragments because of the hostility of the surrounding areas. More study is needed to fully understand how primates move through landscapes with fragmented habitats surrounded by apparently hostile **9**_____, such as grassland.

This study highlights the importance of habitat fragment area over isolation in maintaining stable **10**_____ over time. Conservation efforts should be placed on maintaining or increasing habitat amount. However, for some species, such as common brown lemurs, the impact of isolation is a larger consideration. Fortunately, increasing habitat amount has the secondary effect of reducing isolation in fragmented landscapes.

Lichens Are Bioindicators (8)

Directions: Fill in a word that best completes each sentence.

Lichens, complex symbiotic organisms, which appear as low-growing crust-like plants on rocks, walls, and trees, can live most places on earth. Lichens are found on in the tropics, the Antarctic, the Sahara, and Siberia. Surprisingly, there are some places where even **1**_____ struggle to survive. Since the industrial age in the 18th century, humans have produced numerous sources of air pollution. Air **2**_____ has influenced the health and stability of all ecosystems. Although lichens can survive harsh environmental conditions, many lichen species are highly sensitive to **3**_____ pollution.

In the 19th century, independent observations from England, Munich, and Paris recorded the disappearance of lichens from urban areas. Lichens' sensitivity to air pollution is related to their biology: they take almost all their nutrients and water directly from the atmosphere. They lack stomata and cuticles, so pollutants can be absorbed over their entire surface and they have little biological control over gas exchange. Moreover, they have no deciduous parts to shed after the exposure to a **4**_____, so they can't excrete the toxins.

In 2012 around 7 million people died because of air pollution exposure. Air pollution is now the world's largest single environmental health risk. Lichens can be used as sensitive bioindicators and bio-monitors of air quality to aid in investigating and monitoring air pollution problems. Bioindicators indicate the presence of the pollutants. Bio-monitors are used for experiments measuring the physiological responses to atmospheric pollution over time and providing additional information about the amount and intensity of the exposure. Bioindicators and **5**_____ can give us a biologically-relevant response of an organism to a mixture of pollutants from urban and industrial sources.

Lichens can be used as air **6**_____ detectors. There is biodiversity among lichens, meaning that there is variability in pollutant sensitivities among lichen species. Lichen biodiversity decreases as air pollution levels rise. Severe pollution results in a "lichen desert" – the complete disappearance of lichens, like in the urban areas of the **7**_____ century. Lichens can also be used as **8**_____ or bio-monitors by noting the damage they sustain from pollution. Before pollution causes lichens to die off in an area, they are first placed under biochemical stress. This causes damage that can be monitored by measuring the lichen chlorophyll concentration, secondary metabolites, and the permeability of their cell membranes. Also, since lichens bioaccumulate airborne substances, the build-up of pollutants in lichens can be measured. By measuring the amounts of these substances in lichens, we can deduce the concentrations of these **9**_____ in the ambient air over time.

Lichens get overlooked by most people, but they tell an incredible story of cooperation and persistence. If humans can harness the sensitivity of lichens to pollution, we may be able make our environment cleaner. Keeping pollution levels lower will preserve the world's lichens, as well as human **10**_____.

Lichens Can Be Made of Three Organisms, Not Just Two (9)

Directions: Fill in a word that best completes each sentence.

Lichens can be found all around us: covering the tree by the side of the road or growing on an old building. Previously, lichen species were thought to be a symbiosis of two organisms: algae or cyanobacteria living inside structures made by a single type of fungus.

A new discovery was made by researchers, who were trying (unsuccessfully) to grow lichens in their lab. DNA tests of revealed that two, not just one, types of fungus form a symbiotic relationship with either **1**_____ or cyanobacteria to form a lichen.

Toby Spribille, a postdoctoral researcher at the University of Montana and first author of the study, explained that scientists and natural historians have looked at **2**_____ through microscopes since the 1800s, but his research team was lucky enough to discover a third organism that was there the whole time.

The researchers found this **3**_____ organism, a yeast, when they extended the amount of fungi they were testing for. Instead of staying with the ascomycete fungi that are the known partner in lichens, the researchers tested for basidiomycetes, a separate phylum of **4**_____. And they got a hit.

For years, scientists knew the roles of the symbiotic partners. The ascomycete fungi provide shelter, while the photosynthetic **5**_____, which could be algae or **6**_____, produces food from the sun. It is now hypothesized that this third **7**_____, yeast, protects the organisms, like through the production of toxic vulpinic acid, which was found in one type of yellow-colored lichen.

Once the scientists found the yeast, they tested other specimens from the same collection area, sites in western Montana, and then started testing lichens from all over, like species from Northern Europe. They found that 52 separate genera, on six different continents, also contained **8**_____.

The researchers, surprised by their findings, believe that this just goes to show how more research is always needed. "This discovery overturns our longstanding assumptions about the best-studied symbiotic **9**_____ on the planet," said coauthor M. Catherine Aime, professor of botany and plant pathology at Purdue University in a press release.

"These yeasts comprise a whole lineage that no one knew existed, and yet they are in a variety of lichens on every continent as a third **10**_____ partner. This is an excellent example of how things can be hidden right under our eyes and why it is crucial that we keep studying the microbial world."

Meet the Lichens: Symbionts and Pioneers (10)

Directions: Fill in a word that best completes each sentence.

Lichens appear as crust-like, low growing plants that are typically found on rocks, walls, and trees. In fact, lichens are actually a community of fungi and algae or cyanobacteria. The fungi provide protection from drying out and shield against too much sunshine, while algae or **1**_____ use photosynthesis to produce sugars for both partners.

People thought lichens were just one organism until the invention of the microscope. In 1867, using a **2**_____, Simon Schwendener identified lichens as two organisms – fungi and algae - living supportively side by side. The word "symbiosis" was first used in biology in 1877 to describe a lichen as a close, mutually-beneficial interaction between organisms.

After Schwendener's discovery, it was shown that, in some lichen species, cyanobacteria can replace the algal role as a partner or that there can be both algae and cyanobacteria in living symbiotically with a fungus. Additionally, recent discoveries show that bacteria found on lichens can have an important role in the **3**_____. Still another finding was a discovery led by Toby Spribille in 2016. He found that there is another necessary partner composing a lichen: a form of unicellular fungus which is a symbiotic **4**_____ in almost all lichens, but whose role is still unknown. Scientists are still working to understand the complexity of the organisms known as **5**_____.

Lichens are amazingly diverse in their form, from the dry crustose "skin" you see on rocks, to the coralline-looking lichens in sand and the beard-like lichens hanging from trees. Symbiotic partnerships like this were thriving even before vascular plants evolved and have originated independently several times, which makes lichens an incredibly diverse group for studying. Such a successful cooperation between lichen partners has enabled them to inhabit and flourish in almost all the various habitats on Earth. So **6**_____ are lichens that some are even cryptoendolithic, meaning they live inside rocks.

Lichens are considered pioneers because they are self-sustaining and have few ecological requirements. They need only moisture and sunshine to survive, and this need not even be present throughout the year. Over time, lichens are also able to prepare life conditions necessary to support other organisms. They prepare **7**_____ through the biodeterioration of rocks. They also prepare landscapes for other **8**_____ by accumulating different nutrient elements (i.e. nitrogen, sulphur, phosphorus), thereby increasing availability of these elements for other organisms. Furthermore, after they decompose, they leave the soil enriched with these **9**_____ and other particles trapped from the air. Thus, colonization of a **10-**_____ by lichens is an important first step in the succession of barren locations and areas devastated by disasters like volcanic eruptions.

Taking Manatees Off the Endangered Species List Doesn't Mean We Should Stop Protecting Them (11)

Directions: Fill in a word that best completes each sentence.

Manatees have been on the endangered species list since 1972, but in the last few years their numbers have increased. Though many manatees still get injured in boating accidents or die in cold snaps, their population has more than quadrupled since they were first legally protected.

In 2007 the Fish and Wildlife Service recommended a change in the manatee's status because of its increasing **1**_____, but no action was taken. In 2012, however, the Pacific Legal Foundation petitioned the Fish and Wildlife Service to remove the manatee's **2**_____ status.

The Pacific **3**_____ Foundation isn't an environmental advocacy group. It is a property rights group. The group has been to the Supreme Court multiple times to defend property owners from environmental regulations. This time, they're representing the Save Crystal River organization, a local group of Floridians who want to enjoy the natural riverway they live around. They contend that the restrictions put on the river—rules against touching manatees and regulations on speedboats, for example—are unnecessary.

The Fish and **4**_____ Service is now considering a downgrade in the manatee's **5**_____ because property owners have asked them to, not because of overwhelming evidence of **6**_____ thriving. Most of the Floridian public who voiced their opinions were against a status change. The Save the Manatee Club is against removing the endangered status. They contend that the focus should be on future protections for manatees, not just the current population numbers.

Manatees sometimes die near water treatment plants because the warm areas the plants create aren't as resistant to cold snaps as natural springs are. The treatment plants often provide a safe place for manatees to winter, but the manatees would still be in danger if Florida power companies got rid of those plants. Future uncertainties like these must be considered before removing manatees as an endangered **7**_____.

Even if manatees are taken off the endangered species **8**_____, they'd still have protection under the Endangered Species Act and the Marine Mammal Protection Act. Threatened species can still have all the protections afforded to endangered species, they're just not guaranteed them. The **9**_____ and Wildlife Service evaluates each species to determine which protections it needs. State agencies can "take" threatened species while trying to conserve the animals—that means they're legally allowed to kill, wound, trap, or move a manatee if it's **10**_____, but not if it's endangered. Therefore, the shift in status could be in name only, but it "opens the door" for removing protections that manatees might need.

Migratory Animals Carry More Parasites (12)

Directions: Fill in a word that best completes each sentence.

Every year, billions of animals migrate across the globe, carrying parasites with them and encountering parasites through their travels. A team of researchers at the University of Georgia's Odum School of Ecology discovered that animals known to **1**_____ long distances are infected by a greater number of parasite species than animals that do not migrate. The researchers used parasite records from a database called the Global Mammal Parasite Database 2.0, and focused on ungulates, also known as hooved animals, a group that includes deer, wildebeests, caribou, antelope and gazelles.

Claire Teitelbaum and Shan Huang, both of UGA, spearheaded the project. The most common reason why animals that migrate long **2**_____ are infected by more **3**_____ is that they travel far and, as they do, they pick up more parasites as they go. Migrants can also experience environments that better support parasite transmission year-round through their annual movement cycles.

Previous research on parasites in migratory animals has focused mostly on birds and butterflies, so the team chose to focus on hooved **4**_____ instead. Teitelbaum grouped ungulate species based on whether they migrate seasonally, move unpredictably, or not at all. This allowed the team to compare the number of parasite species in nomadic, resident and migratory species. "We wanted to look at a slightly less well-represented group of mobile animals," said Richard Hall, an assistant professor in the school of ecology who helped oversee the project. "We also wanted to compare across host and parasite species and see whether there were any general patterns."

The study's findings can help researchers understand whether migrants are at greater risk from parasites, discover better ways to protect **5**_____ whose populations are declining, and learn more about parasites living in cattle and other animals that are important food sources for people. "Sheep, cattle, goats, pigs, these are also **6**_____ animals," said Sonia Altizer, also of UGA. "Wild animals can share parasites and infectious diseases with their domesticated animal relatives, and if these migrants are moving around through diverse landscapes, it's important that we understand how much they are exchanging parasites and **7**_____ diseases with livestock and **8**_____ animals."

Hall and Altizer previously thought of migration was a way that animals avoid parasites. They were expecting that fewer migrating animals would be infected, because they leave behind **9**_____ where they might pick up parasites. They also thought that migrating animals with parasites might die during migration. Instead, they found the opposite. The fact that migrants have more parasites does not mean they are sicker than non-migrants. "The number of parasites that they have doesn't necessarily mean that they are suffering more from disease," Teitelbaum said. "They could be infected by parasites that are less **10**_____. From a conservation perspective, migratory animals are in bad shape because of climate change."

Rabbit Relatives Reel from Climate Change (13)

Directions: Fill in a word that best completes each sentence.

Pikas, a hamster-size rabbit relative, have disappeared from a 64-square-mile plot in the northern Sierra Nevada—and climate change is a likely culprit. Up in California's High Sierra, above the dense pine forests, rocky habitats reign. And if you look carefully among the boulders, you might see a pika—a **1**_____ relative the size of a **2**_____, with round ears and big eyes.

"Hikers often see them with little bouquets of wildflowers sticking out of their **3**_____," said Joseph Stewart, a conservation biologist at the University of California Santa Cruz. "Maybe I'm a little biased, other people tell me they're very cute. I find them majestic. I feel like they're the lords of the mountain."

Stewart is also a skilled spotter of pikas. "You could say I've got a little bit of experience with that." Which makes it odd that in five years of surveys—of 64 square miles of high mountain rocky habitats near Lake Tahoe—he found no **4**_____ at all in an area littered with decades-old pika droppings.

It is a habitat in the center, rather than the edge, of historic pika territory. Stewart suspects the most likely culprit for this local extinction is rising temperatures, due to **5**_____ change.

The temperature is predicted to **6**_____. "By 2050 we expect there's going to be a 97 percent decline in the area of climatically suitable habitat in the greater Tahoe area."

Pikas can still be seen elsewhere in Tahoe, and in other parts of the Sierras. Especially further south, where higher mountains allow pikas to escape rising **7**_____. Still, Stewart worries about what will eventually become of pikas.

"We're depriving future generations of the ability to see this critter. And pikas have been around **8**_____ than humans have been around. What right do we have to cause **9**_____ and all the other species that are vulnerable to climate **10**_____ to go extinct?"

Resurrecting Extinct Animals Might Do More Harm Than Good (14)

Directions: Fill in a word that best completes each sentence.

Land use changes and newly introduced predators such as cats and ferrets led to the extinction of New Zealand's laughing owl in 1914. Genetic engineering techniques could bring **1**_____ owls and other extinct species back. Advocates of the practice believe it would be a way to fix some of the damage that humans have caused to the environment. According to a paper published in *Nature Ecology and Evolution*, however, a de-extinction program that provides "environmental justice" could come at a cost to other species that need help. In fact, resurrecting extinct **2**_____ could lead to a net loss of biodiversity.

While de-extinction opponents have often cited small conservation budgets as an argument against resurrection, Bennett's team managed to put a price tag on the rebirth of a few specific species. Once a **3**_____ is brought back from the dead, their populations will be small, and they'll need the same types of protections that endangered species require. The cost of managing the new-old species would fall on either governments or private institutions. Bennett's team analyzed how either of those scenarios would affect the conservation of species that already exist, and have never been **4**_____.

Bennett made estimates for 11 extinct New Zealand species, and 5 extinct species from New South Wales. He often found that resurrecting one extinct species led to the loss of many other species due to competition for financing. The reason the (currently) extinct species would be more expensive to conserve is because their threats—habitat loss or predators, for example—were more difficult to mitigate. "Maybe that's why some of them went extinct," says Bennett. "Another reason for their higher **5**_____ is that we considered captive breeding and re-introduction, which are not necessary for most existing species."

Ben Novak, an ecologist working to restore the extinct passenger pigeon, cautions against extending the predictions from New Zealand and New South Wales across the entire globe—the impact of de-extinction is going to vary from location to **6**_____ and species to species.

In a commentary that accompanies the paper in *Nature Ecology & Evolution*, Ronald Sandler, an ethics professor at Northeastern University, notes that the most prominent argument in favor of de-extinction isn't focused on a cost-effective conservation tool, rather it focuses on the moral arguments for **7**_____ that no study can quantify. De-extinction is about "**8**_____ justice", recreating lost values, and revising conservation paradigms.

Bennett agrees that there are compelling reasons to work on de-extinction. For example, it could lead to techniques that help to **9**_____ existing species by adding genetic diversity into a population very quickly. "If indeed de-extinction is going to be used as conservation **10**_____, then people need to have a very sober look about what those resources would do for existing species," says Bennett.

Russian Cuckoos Are Taking Over Alaska (15)

Directions: Fill in a word that best completes each sentence.

Both the common and oriental cuckoos are "brood parasites," which means they lay eggs in the nests of other **1**_____ and rely on those unwitting foster parents to care for their young. The cuckoo babies usually hatch first, then proceed to murder rightful nestlings and chuck eggs out of the **2**_____. Birds that haven't evolved to detect cuckoos will take care of their solo youngster, devoting all their time and attention to it. The result: no new nestlings from the host parents. If this happens in enough nests, it can impact the survival of a given species. This would be especially worrisome in Alaska, where many of the native birds are rare and specialized.

In the cuckoo's historic range, native species have gotten smart to the cuckoo's ways. They cope with this potential incursion by hiding their nests. Birds that are native to the cuckoo's **3**_____ also toss out cuckoo eggs before they have time to **4**_____.

"It looks like cuckoos are ready to invade North America," says University of Illinois ornithologist Mark Hauber. In the past decade, Hauber says, there has been an increase in cuckoo sightings in both Siberia and Alaska, something likely related to climate change. This poses a huge issue for warblers, buntings, and wagtails in Alaska, as they tend to be specialized and defenseless against cuckoos' strategies.

Hauber and his multi-university team of colleagues conducted a study where they placed more than 100 3-D printed cuckoo eggs in the nests of birds in Siberia and in **5**_____. In Siberia, although the sites were all outside the usual cuckoo nesting range, 14 of the 22 eggs were rejected, suggesting that "Siberian birds had strong anti-parasite responses." In Alaska, however, only one among the 96 eggs planted was **6**_____, by a Red-throated pipit. The other test subjects accepted their **7**_____, not seeming to mind that it differed from their own in color and size.

If those had been real cuckoo eggs, the parents would likely have squandered their breeding season taking care of a demanding chick, rather than breeding again—even if the first nest included a cuckoo that did not survive. It was surprising to see "no rejection whatsoever from Alaskan birds," Hauber says.

Because the **8**_____ pose a threat to native **9**_____, more study is needed to figure out how they are moving during breeding season and which species they are targeting, says Hauber. He'd like to see researchers fit some cuckoos with radio trackers and begin this work.

"While an invasion may spell trouble for Alaskan songbirds, the evolutionary implications are intriguing," says Mary Caswell Stoddard of Princeton University, who was not involved in study. "Will cuckoos evolve better egg mimicry to deceive these new hosts? Will **10**_____ begin to evolve defenses, like mobbing cuckoos or developing more recognizable patterns on their own eggs? A cuckoo invasion into Alaska would open up tons of new questions and research possibilities."

Sensors Applied to Plant Leaves Warn of Water Shortage (16)

Directions: Fill in a word that best completes each sentence.

Forgot to water that plant on your desk again? The plant may soon be able to send out a distress signal. MIT engineers have created sensors that can be printed onto plant leaves and reveal when the plants are experiencing a water shortage. This kind of technology could not only save neglected houseplants but, more importantly, give farmers an early warning when their crops are in danger, says Michael Strano, Professor of Chemical Engineering at MIT and a senior author of the new study, which appears in the journal *Lab on a Chip*.

Strano says, "It's hard to get this information any other way. You can put sensors into the soil, or you can do satellite imaging and mapping, but you never really know what a particular **1**_____ is detecting as the water potential." When soil dries out, plants slow down their growth, reduce photosynthetic activity, and suffer damage to their tissues. Some plants begin to wilt, but others show no visible signs of trouble until they have already experienced significant harm.

The new MIT sensor takes advantage of plants' stomata -- small pores in the surface of a leaf that allow water to evaporate. As water evaporates from the **2**_____, water pressure in the plant falls, allowing it to draw **3**_____ up from the soil through a process called transpiration. Plant biologists know that **4**_____ open when exposed to light and close in darkness, but the dynamics of this opening and **5**_____ have been little studied because there hasn't been a good way to directly measure them in real time. "People already knew that stomata respond to light, to carbon dioxide concentration, to drought, but now we have been able to monitor it continuously," says Volodymyr Koman, an MIT postdoc, and lead author of the paper. "Previous methods were unable to produce this kind of information."

The MIT researchers created their **6**_____ using an ink made of tiny hollow electricity conducting tubes of carbon dissolved in an organic compound. This ink gets printed across a pore to create an electronic circuit. When the pore is closed, the circuit is closed, and the current can be measured by connecting the circuit to a device called a multimeter. When the pore opens, the **7**_____ is broken and the current stops flowing. This **8**_____ and closing allows the researchers to take measurements and observe the stomata's reactions under normal and dry conditions. The researchers found that they can detect, within two days, when a plant is experiencing water stress. The plant used in the study, the peace lily, was chosen for its large stomata.

The MIT team is now working on a new way to apply the **9**_____ circuits by placing a sticker on the leaf surface. In addition to large-scale agricultural producers, gardeners and urban farmers may be interested in such a device. "It could have implications for farming, especially with climate change, where you have **10**_____ shortages and changes in environmental temperatures," Koman says.

Springtime Now Arrives Earlier for Birds (17)

Directions: Fill in a word that best completes each sentence.

In 1869, the Smiley family purchased 8,000 acres of land about 100 miles north of New York City. It later became the Mohonk Preserve, which attracts droves of visitors to its thick forests and rugged crags—popular among rock climbers.

The Mohonk Preserve has a long scientific legacy. In the 1930s Dan Smiley, a descendant of the original owners, began keeping track of the plants and animals that lived in the **1**_____. He wrote thorough notes on the backs of menus from the Mohonk Mountain House, a nearby resort also owned by his family. He would take these old **2**_____ because he wanted to save paper. He was an environmentalist. He cut up old menus into squares. On the back of his "index cards" there are parts of the menu from the 1920s and 1930s, so you can see what was served for **3**_____ that night.

Smiley's efforts produced a rare long-term data set of observations. It's ideal for studying the impacts of climate change, which often play out over the course of many decades and cause subtle changes in the timing of natural processes. For instance, one set of observations shows that songbirds are migrating north earlier and earlier in the spring.

Megan Napoli, a research ecologist with the Mohonk Preserve in New York, explains that it's important that the birds arrive at the proper time in the **4**_____ because they need to time their arrival with the insect emergence. They need to be here to establish their territories and their nesting sites. Then they lay their **5**_____. The hatching of the egg needs to be timed to coordinate with when the **6**_____ are most abundant.

Napoli has begun analyzing roughly 76,000 observations of songbird migration dates collected by Smiley and his team to see if they too show that **7**_____ change has altered the timing of **8**_____. Her preliminary results suggest that they do. Napoli found that short-distance migrants that spend their winters in the southern U.S. now arrive an average of 11 days earlier than they did in the 1930s. Long-distance migrants that overwinter in the tropics arrive roughly a week **9**_____. Napoli presented her results at a recent Ecological Society of America meeting in Portland, Oregon.

As in previous studies, Napoli also found a correlation between early arrivals and rising spring temperatures at Mohonk, which the Smiley family has been tracking since 1896. But she says there are still more questions about how and why the birds are migrating earlier, and Smiley's data may hold more clues.

Meanwhile, who knows how many other long-term, personal **10**_____ collections like Smiley's are out there, waiting to be discovered. They could help bolster official attempts to track the planet's changes.

Tick Paralysis (18)

Directions: Fill in a word that best completes each sentence.

A five-year-old girl in Mississippi temporarily lost the ability to walk after she developed "tick paralysis," a rare condition caused by tick bites. Tick **1**_____ is the only tick-borne disease that is not caused by an infectious organism. The girl's mother first noticed something was wrong when her daughter had trouble getting up to go to daycare.

"As soon as her feet hit the floor, she fell," the mother told a local news outlet in **2**_____. "She would try to stand and walk but would continue to fall." At first, she thought her daughter's legs were asleep, but she later found a tick on the girl's scalp while she was brushing her **3**_____. After removing the **4**_____ and arriving at the emergency room, the five-year-old child was diagnosed with tick paralysis.

Tick paralysis occurs when an engorged and gravid (egg-laden) female tick produces a neurotoxin in its salivary glands and transmits it to its host during feeding. Experiments have indicated that the greatest amount of toxin is produced between the fifth and seventh day of attachment, although the timing may vary depending on the species of tick. Most North American cases of tick paralysis occur from April to June, when adult *Dermacentor* ticks emerge from hibernation and actively seek hosts.

Unlike Lyme disease, which involves the transmission of the *Borrelia burgdorferi* bacterium, which causes ongoing symptoms in its **5**_____ long after the offending tick is gone, tick paralysis is chemically induced by the tick and therefore usually only continues in its presence. Once the tick is removed, **6**_____ usually diminish rapidly. However, in some cases, more profound paralysis can develop before the host is aware of the tick's **7**_____.

Symptoms of tick paralysis usually include an unsteady gait, muscle weakness and eventually, breathing difficulties. The paralysis is "ascending," which means it starts in the lower body and moves **8**_____. The condition can also lead to flu-like symptoms such as muscles aches and tiredness. Tick paralysis is of concern in domestic animals and livestock in the United States. Human cases are rare and usually occur in children under the age of 10. Most previous cases of tick paralysis have been reported in female children, according to a 2012 report on the condition. Girls often have longer hair than boys, which ticks can attach to and hide in, increasing the risk of tick paralysis.

It is important for parents to check their children for ticks. The five-**9**_____-old Mississippi **10**_____ fully recovered from the tick paralysis after a couple of days.

Evolution

How Did 3D Vision Develop? (1)

Directions: Fill in a word that best completes each sentence.

The human retinas, the parts of each eye with light sensitive cells that help us see, have nerve fibers which extend directly back to the rear of the brain. That is, many of the fibers from the right eye travel back to the rear of the right brain hemisphere; many of the fibers from the left eye travel back to the rear of the **1**_____ brain **2**_____. The image is formed in the visual cortex at the back of the human brain.

Isaac Newton observed that the way many human optic nerves stay on the same side of the brain from which they originate is different from other animals. In most animals, almost all optic nerve fibers cross to the opposite side of the brain. Newton's observations led to a widely accepted concept that the fewer optic nerve fibers that cross to the **3**_____ side of the brain, the larger the binocular field of view and the better the depth perception (stereopsis) of the organism. When the eyes perceive an object from slightly different angles, the brain is better able to estimate distance. The general assumption among researchers has been that the arrangement of **4**_____ fibers in primates (including humans) is primarily is intended to create accurate depth perception.

A new proposal, termed the eye-forelimb hypothesis, challenges the idea that the purpose of ipsilateral (same side) projection of nerve fibers was to provide stereopsis (depth **5**_____). The eye-forelimb hypothesis suggests that retina and surrounding structures are constructed so animals (including humans) can steer the forelimbs (hands, claws, wings or fins). Nerve cells that control right hand movement, that receive sensory impressions from that hand, and that receive visual information about that hand, all end up in the same brain hemisphere. The opposite applies to the left hand. These conditions aid the eyes in being of service to the forelimb; visual information from the **6**_____ will reach the appropriate hemisphere.

Another claim of the stereopsis hypothesis is that predatory animals generally have frontally-placed eyes to enable them to estimate the **7**_____ to their prey, while animals preyed-upon have laterally-positioned eyes, which allow them to scan their surroundings and detect the enemy in time. This logic is flawed because most predatory animals may also become prey to other predators, and many predatory animals, for example the crocodile, have laterally situated eyes.

Yet, the eye-forelimb hypothesis is also imperfect. It is not supported when we observe owls. Owls have frontal eyes and primate-like visual systems. As raptors, they don't use their wings (forelimbs) to hunt, they use their feet (hindlimbs). The **8**_____-forelimb hypothesis incorporates Newton's idea of **9**_____ perception. Although all scientific questions about the evolution of 3D **10**_____ have not been solved, the eye-forelimb hypothesis may provide us with a better understanding of how humans' ability to estimate distance developed.

The Battle of the Sexes Can Show Us How to Live Longer (2)

Directions: Fill in a word that best completes each sentence.

There are important evolutionary differences between men and women, and understanding them can help us live longer, healthier lives, scientists say. These differences might commonly be called "the battle of the sexes", but the scientific term is sexual antagonism. Sexual **1**_____ includes the evolutionary genetic variants that are good for one sex but bad for the other.

One example of sexual antagonism is longevity. No matter how advanced medicine becomes, women, on average, still live longer than men. Dr Alexei Maklakov, of Uppsala University in Sweden, explains that males and females have different reproductive strategies. This means that the strategy that the human population of males has used to survive is very different than the **2**_____ that the human **3**_____ of females has used to survive during that same period. There is a trade-off between reproduction and longevity.

The number of people aged 65 years or over are projected to increase in the EU by ten percent in the coming years. As the **4**_____ population ages, it important to study this subject. "There is a massive research effort underway to find treatments that can improve the human condition in late life," said Dr Maklakov. "However, because of the fundamental biological differences between the sexes, men and **5**_____ are likely to respond differently to the same treatments."

As part of the EU-funded AGINGSEXDIFF project, Dr Maklakov has done tests on roundworms and fruit flies to look at the factors which affect longevity. The project also analyzed human data to help explain the variation in longevity between species, populations, sexes and individuals.

In female beetles, for example, longevity is positively associated with fitness, meaning longer-lived females produce more offspring. In male beetles it's a different story: successful males invest a lot in early-life reproduction and die young. This same principle explains the trade-**6**_____ between reproduction and **7**_____ that is seen in the human population. In the past, women paid a higher cost than **8**_____ in terms of longevity for having **9**_____: they often died in childbirth. A historical shift to smaller families has increased longevity among women but not among men, Dr Maklakov explains.

Lifespan- and healthspan-extending treatments in humans will likely always affect men and women differently. Dr Maklakov believes that we need more research on sex-specific effects of new treatments. Similarly, Dr Jessica Abbott at Lund University in Sweden, who has been working on a related project, notes that "Lots of diseases with a genetic component have different prevalence in men and women. My colleagues and I have speculated that this might be partially due to **10**_____ antagonism,' she said. 'The better we understand sexual antagonism as a phenomenon, the more we might be able to understand the reasons for these differences between men and women in disease and ageing."

Are Cities Affecting Evolution? (3)

Directions: Fill in a word that best completes each sentence.

"As we build cities, we have little understanding of how they are influencing organisms that live there," says Marc Johnson, a director of the University of Toronto's Centre for Urban Environments. Johnson says that it's good news that some native species can adapt to **1**_____ because they have important ecological functions in the environment. The bad news, however is that some of these organisms' adaptations might increase the transmission of disease. For example, bedbugs, which used to be scarce two decades ago, have adapted to the insecticides used to keep them from flourishing and now their population has exploded worldwide.

Johnson and Jason Munshi-South, of Fordham University, recently reviewed all existing research studies about urbanization and evolution and synthesized the results. Traditionally, **2**_____ has been thought to be a long-term process driven by the natural environment and interactions between species. Another factor is also at work: humans and the man-made **3**_____.

Loss of habitat and urban barriers (roads, buildings, etc.) present challenges to all kinds of species and some may adapt in undesirable ways. This research identifies evolutionary impacts on species as diverse as the common blackbird in Europe to clovers in North America. For example, populations of white-footed mice in New York City became differentiated from each other due to their isolation in various city parks.

The study, published in the journal, *Science*, is a "wake-up call for the public, governments and other scientists." The researchers suggest that we need to think carefully about how we're altering our environment in unintended ways when we build cities, influencing the evolution of **4**_____ that may, in turn, influence our lives. Many organisms, such as rats, urban lizards, cockroaches, pigeons and bedbugs, have evolved to depend on humans. Johnson and Munshi-South suggest that when we're planning cities, we need to think about the impact our designs have on native species and whether we can design them to "be kinder to ourselves and the environment," considering ways to conserve **5**_____ species and lessen the number of disease-carrying pests.

There are now mosquitoes, for example, that have evolved to live in the London Underground stations and adapted so they no longer need to feed on blood to produce eggs. They also have no need to be dormant during winter. Unfortunately, these **6**_____ can carry several **7**_____ and are now found in New York City, Chicago and Los Angeles. Healthcare systems need to adapt in response.

Given that **8**_____ are evolving rapidly in response to urbanization, **9**_____ have become a classroom where people can see examples of evolution. **10**_____ evolution can be used as a tool to educate city dwellers and others about the importance of evolutionary biology. "People who don't believe in evolution need not go further than their backyards to see evidence of it," Johnson said.

When Dinosaurs Took Flight (4)

Directions: Fill in a word that best completes each sentence.

Differing skull structures of ancient reptiles are the reason why scientists classified them into different groups: synapsids and the sauropsids. The synapsids gave rise to the mammals, whereas the **1**_____ gave rise to all modern reptiles and birds. Dinosaurs evolved from their sauropsid ancestors around 230 million years ago (mya), and became the dominant animals on land.

Although there is some debate over the dinosaur family tree, we know that birds evolved from a group called the theropods. Many theropod fossils have both dinosaur- and bird-like features. The theropods were almost all bipedal (walked on two legs). Because of this, their arms were able to be used for purposes other than support. The arms of birds' ancestors slowly evolved adaptations for flight.

Bird wings differ from the vertebrate arms in two ways: wings have three digits (fingers), instead of five, and **2**_____ have feathers. Flight feathers contain a central, hollow shaft ('rachis'), and many 'barbs' projecting from the rachis. These barbs also have their own little projections called 'barbules' which cross-link the barbs, allowing the feathers to maintain their rigid structure, which is essential for flight.

In addition to **3**_____ feathers, birds also have fluffy 'down' feathers. Down **4**_____ cover chicks, and the belly and thighs of adult birds. They provide insulation. The barbules on **5**_____ feathers don't cross-link with each other. This means the barbs are free to move around, and the feather has a much less rigid structure.

Birds also have even simpler feathers called 'bristles' around the beak and eyes. These bristles, which are similar in appearance and function to whiskers on mammals, are each composed of a rachis with few, small barbs projecting from them. Scientists think that bristles appeared first for sensing movement. Then, later in evolution, these shafts developed more barbs, which would help provide **6**_____, and these down feathers then spread to other parts of the body. Only later did cross-linking barbules evolve on the wing feathers, which would help the wings remain rigid and strong. Feathers may have evolved for functions other than **7**_____, such mating, aggression displays, or sheltering young.

It is also possible that wings could have been used by early birds climb up into the branches of trees to escape predators. For example, chicks of the *Alectoris chukar*, a type of pheasant, can't fly. They can, however, flap their wings as they run up steep surfaces (like tree trunks) to help them climb. Perhaps the ancestors of **8**_____ evolved a similar strategy. Once in the **9**_____ they could glide back to ground or flap their wings to slow their fall. Over time, these behaviors (and the wings themselves) evolved to allow proper flight.

Around 65 mya there was a mass extinction event which wiped out all **10**_____ – except one group: birds. It's unclear why only these small, feathery dinosaurs survived. Whatever the reason, birds diversified and spread all over the world.

Earth's Orbital Changes Have Influenced Climate, Life Forms For At Least 215 Million Years (5)

Directions: Fill in a word that best completes each sentence.

Every 405,000 years, gravitational tugs from Jupiter and Venus slightly elongate Earth's orbit, an amazingly consistent pattern that has influenced our planet's climate for at least 215 million years and allows scientists to more precisely date geological events like the spread of dinosaurs, according to a Rutgers-led study. The findings have been published online in the Proceedings of the National Academy of Sciences.

"It's an astonishing result because this long cycle, which had been predicted from planetary motions through about 50 million years ago, has been confirmed through at least 215 million **1**_____ ago," said lead author Dennis V. Kent, of Rutgers University-New Brunswick. "Scientists can now link changes in the climate, environment, dinosaurs, mammals and fossils around the world to this 405,000-year cycle in a very precise way."

The scientists linked reversals in the Earth's magnetic field -- when compasses point south instead of **2**_____ and vice versa -- to sediments with and without zircons (minerals with uranium that allow radioactive dating) as well as to climate cycles. "The climate **3**_____ are directly related to how the Earth orbits the sun and slight variations in sunlight reaching **4**_____ lead to climate and ecological changes," said Kent. "The Earth's **5**_____ changes from close to perfectly circular to about 5 percent elongated especially every 405,000 years."

The scientists studied the long-term record of reversals in the Earth's **6**_____ field in sediments in the Newark basin, a prehistoric lake that spanned most of New Jersey, and in sediments with volcanic detritus including zircons in the Chinle Formation in Petrified Forest National Park in Arizona. They collected a core of rock from the Triassic Period, some 202 million to 253 million years **7**_____. The core is 2.5 inches in diameter and about 1,700 feet long, Kent said. The results showed that the 405,000-year cycle is the most regular astronomical pattern linked to the Earth's annual turn around the **8**_____, he said.

Prior to this study, dates to accurately time when magnetic **9**_____ reversed were unavailable for 30 million years of the Late Triassic. That's when dinosaurs and mammals appeared, and the Pangea supercontinent broke up. The break-up led to the Atlantic Ocean forming, with the sea-floor spreading as the continents drifted apart, and a mass extinction event that affected dinosaurs at the end of that period, Kent said.

"Developing a very precise **10**_____-scale allows us to say something new about the fossils, including their differences and similarities in wide-ranging areas," he said.

The First Dogs May Have Been Domesticated in Central Asia (6)

Directions: Fill in a word that best completes each sentence.

As many as 47 percent of all American households have a dog. We affectionately call them man's best friend. In fact, *Canis familiaris*, the domestic dog, was the first species to be domesticated by humans from Eurasian gray wolves at least 15,000 years ago. Scientists are unsure where this occurred, however.

A new genomic study, published in the *Proceedings of the National Academy of Sciences*, offers up strong evidence that domesticated **1**_____ originated in Central Asia, around modern-day Nepal and Mongolia. Previous genetic studies of dog lineages have concluded that domestication occurred in Southern China around 16,500 years ago. More conflicting archaeological evidence has been found of **2**_____ dogs in Europe and Siberia. All this contrasting evidence has created a lot of debate.

Canis **3**_____ can be broken down into two main groups: the pure and mixed breeds that are typically household (domestic) pets and a much larger group of free-roaming and breeding "wild" or "village dogs." Despite the wide variation in outward physical appearance, pure breeds are not genetically very diverse, because they come from such small, controlled gene pools. Modern breeds, like the chihuahua and St. Bernard, are only about 200 years old. "Wild" or "village dogs" are more **4**_____ and have much older lineages than the pure **5**_____. This makes them particularly important when studying dog evolution.

Adam Boyko, of Cornell University, and his colleagues analyzed 185,800 genetic markers in 5,392 dogs, including 549 village dogs from 38 countries, making their study the largest ever of worldwide canine genetic diversity. They found that genetic **6**_____ is highest in Central Asia, specifically Nepal and Mongolia. Genetic diversity then fans out like ripples in a pool, with areas like Afghanistan, Egypt, India, and Vietnam—all ringing around the possible center of origin in Central Asia. "It mirrors what we see in humans and how they spread out of East Africa," says Boyko.

Dogs studied in geographic regions farther away, like the South Pacific and Americas were almost exclusively of European origin. This complicates things a bit, but the reason for that is likely the spread of modern Western culture and the appeal of **7**_____ breeds. "Everybody wants a golden retriever," says Boyko. Perhaps one of the reasons village dogs with more indigenous traits have not been overcome by foreign gene flow in Central Asia is because they originated there and have large populations. They also might be better adapted to that environment, notes Boyko.

Boyko's study shows the first strong evidence of where the cradle of dog civilization might lie, which could help in figuring out how they came to be. The populations of "wild" or "village" indigenous dogs, however, are quickly getting overrun by **8**_____ flow from household (**9**_____) dogs. Therefore, time is of the essence in completing further studies to understand how the human relationship with **10**_____ evolved.

How Fish Conquered the Land (7)

Directions: Fill in a word that best completes each sentence.

Vertebrates include: fish, amphibians, reptiles, birds and mammals. They first evolved around 525 million years ago (mya). Early vertebrates were all fish, living in the seas and rivers. It wasn't until 360 mya that vertebrates began to colonize the land. These land animals, called 'tetrapods' (meaning four-footed), gave rise to amphibians, reptiles, birds and mammals. The first tetrapods were like modern amphibians. They had a flat skull, limbs rather than fins, and a neck separating the head from the body.

Archeologists have found fossilized skeletons of an animal which bore similarities to both fish and early tetrapods. The animal, *Tiktaalik roseae*, lived 375 **1**_____ - just 15 million years before the early tetrapods. It had a bone structure between that of lobe-finned fish fins and early tetrapod limbs. *Tiktaalik* **2**_____ is one of several species that shows tetrapods evolved from the lobe-finned **3**_____. Lobe-finned fish are different from most other fish, in that their fins connect to their trunk with a single, thick bone (like tetrapods) instead of having fins that connect with long, thin rays of bone.

It is likely that **4**_____ *roseae* spent most of its time in shallow waters, occasionally using its sturdy fin-limbs to pull itself up onto land. In the following ten million years, other creatures evolved with limbs that have a humerus, radius, ulna, and digits. The fossil record gives us a good idea of how the skeletons became better adapted to life on **5**_____, with a slow transition from sturdy fins to proper limbs.

Life on land requires not only **6**_____, but a way to obtain oxygen. Most fish absorb oxygen into their blood by drawing water in through their mouth and out through their gills. Tetrapods absorb **7**_____ into their bloodstream in their lungs after air first enters and then exits through their mouth and nostrils.

The fossil record from early tetrapods tells us little about the development of lungs, but the study of embryology provides some information. In human embryos, **8**_____ development starts as a single bud which grows out from the gut tube. This bud then branches to form two lungs. The lungs and the gut are connected to the body where the esophagus and trachea meet in the throat.

Many fish possess a swim bladder which develops in a very similar way. The **9**_____ bladder develops as a bud which grows out from the gut tube. Fish use their swim bladder to be buoyant in the water. Some fish swim to the surface and gulp air directly into the swim bladder or expel air from it to reduce buoyancy. In those fish, the swim bladder is a gas-exchange organ like human lungs. Lobe-finned fish use the swim bladder to take up oxygen into their blood for respiration in the same way as tetrapods do!

It is likely that lungs evolved first in lobe-finned fish and some also evolved sturdier fins which allowed them to pull themselves onto land. Locomotion on land and the ability to breath air enabled them to thrive. These fish became **10**_____, which eventually diversified into the amphibians; reptiles; birds and mammals of today.

How Language Evolved (8)

Directions: Fill in a word that best completes each sentence.

Rhythmic movements, such as walking in tandem (at the same pace), produces short intervals of quiet, between foot-falls. In this short break, other environmental sounds are more easily heard. This relative quiet would have been a selective mechanism that helped our ancestors (that could walk in **1**_____) hear predators and choose appropriate actions to survive. As a result, the survival benefit for **2**_____ in tandem would become more common within the species.

Hearing regular pulses, beats, or rhythms causes animal's brains (including humans') to experience the "reward molecule": dopamine. **3**_____ produces a feeling of enjoyment that early peoples may have felt as they clapped, stomped, and howled around a campfire. These early forays into music, is believed by Charles Darwin, and other scholars, to be a necessary precursor in the evolution of language.

Besides the rhythmic **4**_____ of walking (locomotion) and clapping, early peoples worked with their hands and used tools. The Tool Use Sound Theory of Language Evolution suggests that language development has been stimulated by sounds created by tool use. Human ancestors, in mimicking tool use sounds, would create symbolism with those sounds. Once two early humans agreed on a vocalization (noise using the mouth) that symbolized a subject or event, a word would be created.

The sounds of locomotion have benefits beyond language. Sounds created as a side-effect of locomotion are labelled "incidental sounds of **5**_____" or ISOL for short. When non-human primates move, they do so in an irregular, disorderly fashion. When humans began to walk on two legs, one result was more rhythmic and predictable ISOL. This regularity helps individuals to keep pace, move in synchronicity, with a rhythm. Humans walking in synchronicity is similar to how fish swim and birds **6**_____, each with adaptations to their differing ecological niches.

Moving synchronously has many acoustic (sound/hearing) advantages. Theoretical models suggest that fish and birds, swimming and flying in large groups, can use ISOL to navigate in a group. **7**_____ contains information about the neighbor's distance, size, speed, and frequency of movements. If all animals in a group stop swimming, or flapping, at the same time, a noise reduction follows, and fish or birds can eavesdrop on their surroundings. Thus, synchronous movements make it easier to perceive acoustic information. Moreover, ISOL helps synchronize **8**_____ because the fin-beats, wingbeats or footsteps are easily heard by the closest fish, **9**_____ or human and can help each adjust their own speed, distance, and pacing.

Scientists believe that schooling fish may use analogous acoustical mechanisms as a human orchestra to achieve synchronization. The ability to **10**_____ movements and imitate sound is fundamental for musicality. Scientists further believe that bipedal walking led to music, and, later, to language evolution.

Humans Don't Use More Brainpower Than All Other Animals (9)

Directions: Fill in a word that best completes each sentence.

A study published in the *Journal of Human Evolution* comparing the relative brain costs of 22 species found that, when it comes to brainpower, human brains don't use more energy than other animal **1**_____. "We don't have a uniquely expensive brain," said study author Doug Boyer, assistant professor of evolutionary anthropology at Duke University. "This challenges a major dogma in human evolution studies."

Boyer and graduate student Arianna Harrington researched how humans stack up in terms of brain energy uptake compared with seven other mammals. Energy travels to the brain through blood vessels carrying glucose. Therefore, the researchers measured the area of the bony canals that enclose the cranial arteries. These measurements were considered with published estimates of brain glucose uptake and internal skull volume as an indicator of brain size. The researchers were able to show that larger canals enclose **2**_____ that deliver more blood, and thus **3**_____, to the brain.

The researchers then used statistical technique to calculate brain glucose uptake for an additional 15 **4**_____ for which brain costs were unknown. As expected, the researchers found that humans allot proportionally more **5**_____ to their brains than rodents, Old World monkeys, and great apes such as orangutans and chimpanzees. Relative to resting metabolic rate -- the total amount of calories an animal burns each day just to keep breathing, digesting and staying warm -- the human brain demands more than twice as many **6**_____ as the chimpanzee brain, and at least three to five times more **7**_____ than the brains of squirrels, mice and rabbits.

There are, however, other animals that have hungry brains like the human brain. There appears to be little brain-cost difference between a human and a pen-tailed treeshrew, for example. Even the ring-tailed lemur and the tiny quarter-pound pygmy marmoset, the world's smallest monkey, devote as much of their body energy to their brains as we do. Said Boyer, "The metabolic cost of a structure like the brain is mainly dependent on how big it is, and many animals have bigger brain-to-body mass ratios than **8**_____."

The results suggest that the ability to grow a relatively more expensive brain evolved millions of years before humans, when our primate ancestors and their close relatives split from the branch of the **9**_____ family tree that includes rodents and rabbits.

Previous studies calculated the amount of energy needed to fuel a brain based on neuron counts. But because the current study's method for estimating energy use relies on measurements of bone, rather than soft tissue such as neurons, it is now possible to estimate brain energy demand from the fossilized remains of animals that are extinct too, including early human **10**_____. "All you would need to take the measurements is an intact skull and some of the neck vertebrae," Harrington said.

New Species Can Develop in as Little as Two Generations, Galapagos Study Finds (10)

Directions: Fill in a word that best completes each sentence.

Researchers from Princeton University and Uppsala University in Sweden report that a migrant bird who arrived at an island in the Galapagos 36 years ago mated with a member of another species resident on the **1**_____, giving rise to a new **2**_____ that today consists of roughly 30 individuals.

In 1981, a graduate student working on the island of Daphne Major with B. Rosemary and Peter Grant, two scientists from Princeton, noticed a newcomer. The migrant, a male, that sang an unusual song was larger in body and beak size than the three resident species of **3**_____ on the island. "We didn't see him fly in from over the sea, but we noticed him shortly after he arrived. He was so different from the other birds that we knew he did not hatch from an egg on Daphne **4**_____," said Peter Grant.

The researchers took a blood sample and released the bird, which later bred with a resident medium ground finch of the species *Geospiz fortis*, initiating a new lineage. The Grants and their research team followed the new "Big Bird lineage" for six generations, taking blood samples for use in genetic analysis.

In the current study, researchers from Uppsala University analyzed DNA collected from the parent birds and their offspring over the years. The investigators discovered that the original male parent was a large cactus finch of the species *Geospiza conirostris* from Española island, which is about 62 miles to the southeast in the archipelago. The remarkable distance meant that the male finch was not able to return home to mate with a member of his own species and so chose a **5**_____ from among the three species already on **6**_____ Major.

This reproductive isolation is considered a critical step in the development of a new species when two separate species interbreed. Like the original migrant male, the offspring were reproductively isolated because their **7**_____ was unusual and failed to attract females from the resident species. The offspring also differed from the resident species in **8**_____ size and shape, which is a major cue for mate choice. As a result, the offspring mated with members of their own lineage, strengthening the development of the new species.

Researchers previously assumed that the formation of a new species takes a very long time. However, in the Big Bird **9**_____ it happened in just two generations, according to observations made by the Grants in the field. This data is further supported by **10**_____ studies.

It is likely that new lineages like the Big Birds have originated many times during the evolution of Darwin's finches, according to the authors. Most of these lineages have gone extinct, but some may have led to the evolution of contemporary species. The researchers have no indication about the long-term survival of the Big Bird lineage, but it has the potential to succeed. It provides an example of one way in which speciation occurs.

Nose Shape Dependent on Ancestral Climate (11)

Directions: Fill in a word that best completes each sentence.

A new study, published in *PLOS Genetics,* showed that nose shape could be influenced by climate. The researchers, using 3D imaging of live participants, found that the noses of people whose ancestors come from the colder, drier climates of West Africa, South Asia, East Asia, and Northern Europe do tend to have narrower **1**_____ than those from warmer, humid **2**_____, even when accounting for the otherwise random nature of their nose shape. Study lead author Arslan Zaidi, from Pennsylvania State University, believes that the correlation makes sense because noses are the first line of defense against air entering the body as we breathe. "The lungs are very exposed. The **3**_____ coming into your lungs carries cold, dryness, bacteria, pathogens and particles. The internal respiratory tracts lining the nose have mechanisms which capture particles and pathogens and warm and humidify the air that's coming in before it reaches the **4**_____. Colder air is less efficient for retaining moisture, which turns out to be important for trapping those [substances] that if not captured earlier on, could lead to respiratory diseases."

This climate mechanism could be responsible for the evolutionary pressure that drove the differences in nose **5**_____. Although nose shape might seem like an unimportant example of genetic adaptation to the environment, other more prominent examples can play a big role in health today. Skin color and pigmentation, serves as another example of an inherited trait that reflects selective pressures on our ancestors to adapt to their particular climate.

Early dark-skinned individuals evolved with higher amounts of pigmentation in regions close to the equator, since they were exposed to higher amounts of harmful UV radiation from the **6**_____. Meanwhile those populations closer to the Earth's poles developed lighter skin with less **7**_____ to allow more sunlight to be absorbed to produce vitamin D.

Zaidi says that, as the world's population becomes more globalized and people move to different areas on extremely short timescales, our inherited adaptations may serve to harm us, rather than help. For example, dark skinned people living at higher latitudes may need vitamin **8**_____ supplements to compensate for their not being well adapted to environments with less sunlight. Whether it's nose shapes or **9**_____ color, Zaidi says studying how these **10**_____ came about gives us new ways to understand and treat such health problems as they arise today.

Zaidi notes that, despite these differences, it's remarkable how similar we all are. "Traits such as skin pigmentation and nose width have evolved faster than most other traits likely due to natural selection because of exposure to the environment. They are an exception rather than the rule. This is important to mention because often people focus on differences and ignore the similarities."

Archaeologists May Have Discovered One of the Earliest Examples of a 'Crayon' (12)

Directions: Fill in a word that best completes each sentence.

Archaeologists may have discovered one of the earliest examples of a 'crayon' -- possibly used by our ancestors 10,000 years ago for applying color to their animal skins or for artwork. The archaeological site where the **1**_____ was found was near in North Yorkshire, England. This area is known for its low land areas which are covered wholly or partially with water. The terms fen or carr describe these areas of low land where there is a rich peat soil.

The found crayon contains the pigment ochre (pronounced: oker) which can range from light yellow to brown or red. The crayon measures 22mm long and 7mm wide. An ochre pebble was also found, which had a heavily striated surface that had likely been scraped to produce a red pigment powder.

Ochre was an important mineral pigment used by prehistoric hunter-gatherers across the globe. The latest finds suggest that **2**_____ was collected and processed in different ways during the Mesolithic period, a time when early people used stone tools. The ochre crayon and **3**_____ were studied as part of an interdisciplinary collaboration between the Departments of Archaeology and Physics at the University of York, using state-of-the-art techniques to establish their composition.

The artifacts were found at Seamer Carr and Flixton School House. Both sites are situated in a landscape rich in prehistory, including one of the most famous Mesolithic sites in Europe, Star Carr near Scarborough in North Yorkshire. In fact, Star **4**_____ was the site of a 2015 discovery of a pendant, which is the earliest known Mesolithic art in England. Also found at **5**_____ Carr were more than 30 red deer antler headdresses, which may have been used as a disguise in hunting, or during ritual performances by shamans when communicating with animal spirits.

Lead author, Dr Andy Needham from the University of York's Department of **6**_____, said the latest discoveries helped further our understanding of Mesolithic life. He commented: "Color was a very significant part of hunter-gatherer life and ochre gives you a very vibrant **7**_____ color. It is very important in the **8**_____ period and seems to be used in a number of ways. One of the latest objects we have found looks exactly like a crayon; the tip is faceted and has gone from a rounded end to a really sharpened end, suggesting it has been used. It is a very significant object and helps us build a bigger picture of what **9**_____ was like in the area."

The research team say Flixton School House was a key location in the Mesolithic **10**_____ and the crayon and pebble show how the people interacted with the local environment. Both were located in an area already rich in art.

Pigeons Can Read a Little Bit, New Research Shows (13)

Directions: Fill in a word that best completes each sentence.

They might not always know the difference between a straw and a French Fry, but pigeons can recognize words. How they learn to do this could provide insights on the origins of language, according to a study published in Proceedings of the National Academy of Sciences.

The **1**_____ studied were living in a lab in New Zealand where, over a span of two years, they learned to distinguish four-letter English words from nonsense **2**_____. For their training, a computer screen would flash words like "DOWN" or "GAME", and non-words like "TWOR" or "NELD", along with a star symbol. Each time the pigeons made a correct identification — pecking the word if it was a real one, or pecking the **3**_____ symbol beneath a non-word — they were rewarded with some wheat.

After the pigeons built their vocabularies (the star pupil acquired 58 words), the screen began flashing new words that they had never seen **4**_____. Even when faced with new words, the pigeons continued to pick out the **5**_____ words from the non-words with impressive accuracy.

"It appears that the pigeons are paying attention to pairs of letters in the words," explains study lead author Damian Scarf, a lecturer in psychology at the University of Otago, New Zealand. Letters that appear side-by-side are known as bigrams, and some bigrams occur more frequently than others. For example, "TH" is a high-frequency **6**_____, whereas the "CB" combination is far less common. Over time, the pigeons came to pick up on these word properties. Like people, the birds recognized words better if they included common **7**_____ pairs.

It's important to note that these birds were not reading. **8**_____ requires not only the ability to visually recognize words, but also to decode the letter-sound relationships. The pigeons were missing that second half of the equation. But what these birds did manage to learn is remarkable, and it might even explain why humans have an entire brain region devoted to recognizing written **9**_____.

A process called "neuronal recycling" involves brain cells that were once devoted to spotting everyday objects, like rocks or trees, gradually learning to key in to new visuals, like the written word. Some scientists believe this is precisely how ancient **10**_____ first developed reading skills. According to this pigeon study, visual word recognition is not limited to the realm of the primate brain.

Prehistoric Women had Stronger Arms Than Today's Elite Rowing Crews (14)

Directions: Fill in a word that best completes each sentence.

A study comparing the bones of European women that lived during the first 6,000 years of farming with those of modern athletes has shown that the average prehistoric agricultural woman had stronger upper arms than female rowing champions from the 2017 Cambridge University Women's Crew Rowing Team.

Until now, bio-archaeological investigations of past behavior have interpreted women's bones solely through direct comparison to those of men. However, male bones respond to strain in a more dramatic way than female **1**_____. Researchers from the University of Cambridge's Department of Archaeology say this has resulted in the underestimation of the scale of the physical demands borne by women in prehistory. "This is the first study to actually compare prehistoric **2**_____ bones to those of living **3**_____," said Dr Alison Macintosh, lead author of the study published in the journal *Science Advances*.

"It can be easy to forget that bone is a living tissue, one that responds to the rigors we put our bodies through. Physical impact and muscle activity both put strain on bone, called loading. The bone reacts by changing in shape, curvature, thickness and density over time to accommodate repeated **4**_____," said Macintosh. "By analyzing the bone characteristics of living people whose regular physical exertion is known, and comparing them to the characteristics of ancient bones, we can start to interpret the kinds of labor our ancestors were performing in prehistory."

The members of the Cambridge University Women's **5**_____ Rowing **6**_____ were training twice a day and rowing an average of 120km a week at the time. The Neolithic women analyzed in the study (from 7400-7000 years ago) had similar leg bone strength to modern rowers, but their arm bones were 11-16% stronger for their size than the **7**_____, and 30% stronger than typical Cambridge students. The loading of the upper limbs was even more dominant in the study's Bronze Age women (from 4300-3500 years ago), who had 9-13% stronger arm bones than the rowers but 12% weaker **8**_____ bones.

"We can't say specifically what behaviors were causing the bone **9**_____ we found. However, a major activity in early agriculture was converting grain into flour, and this was likely performed by women," said Macintosh. Grain would have been ground by hand for five hours a day. The repetitive arm action used in grinding may have loaded women's **10**_____ bones in a similar way to the laborious back-and-forth motion of rowing. Additionally, women were probably also manually planting, tilling and harvesting crops, fetching food and water for domestic livestock, processing milk and meat, and converting hides and wool into textiles.

Dr Jay Stock, senior study author added: "Our findings suggest that for thousands of years, the rigorous manual labor of women was a crucial driver of early farming economies. The research demonstrates what we can learn about the human past through better understanding of human variation today."

The Rise of the Mammals (15)

Directions: Fill in a word that best completes each sentence.

Early amphibians had limbs and lungs for life on land, but, like those of today, they needed to live near water because that is where they lay their eggs. One group of early **1**_____ began to evolve adaptations which were better suited for terrestrial (land) environments. They produced hard eggs which could be laid on **2**_____, allowing them to live in drier areas. By 320 mya, this group had evolved into the early reptiles.

Those reptiles split into two groups: the sauropsids and the synapsids. The two groups were similar, but with slight differences in skull structure. The sauropsids would give rise to modern reptiles and birds; the synapsids would give rise to mammals. From 320-250 mya, the synapsids were the dominant predators.

The end-Permian extinction caused 95% of all species became extinct, including most synapsids. In their place, one group of sauropsids became the dominant **3**_____: dinosaurs. The few synapsids that survived were small insect-eaters. Scientists believe that they were nocturnal (active at night). This is likely because, during the **4**_____, they would have been eaten (or outcompeted) by dinosaurs. As nocturnal creatures, the sense of sight is weakened. Nocturnal synapsids had to rely on their hearing.

Due to the nocturnal ancestry of mammals, their ears function differently than reptiles. Reptiles have one small inner-ear bone; mammals have three. Three **5**_____ allow mammals to hear higher frequencies than reptiles. Nocturnal **6**_____ would have used this ability to hear their insect prey.

While mammals have improved hearing compared to reptiles, they have poorer color vision. Most reptiles have four color receptors; most mammals have two. It is likely that the other two **7**_____ were lost in evolution because the ability to see a wide range of colors was not useful to nocturnal life.

Characteristics which aid survival are found more frequently in populations. Nocturnal synapsids needed to stay warm to survive at night. Mammals had to eat more food than reptiles to produce their own heat. Mammal cells "lose" a lot of the energy that comes from greater food consumption as heat. The hair on mammalian bodies traps a layer of air around the skin, which helps to keep in the **8**_____.

Three mammal groups appeared around 160 mya: monotremes, marsupials, and placentals. The monotremes, like the platypus, are like reptiles in that they lay eggs rather than give birth to live young. The marsupials, like kangaroos, give birth to underdeveloped **9**_____, which then live in the mother's pouch. The placentals, which include humans, nourish their young inside the womb via a placenta before they are born. The mammals continued as predominantly small, nocturnal creatures until around 65 mya, when the dinosaurs went extinct. After their extinction the mammals were able to increase in size and diversify further. Although highly diverse, mammals still share many traits which can be traced back to their small, nocturnal ancestors: advanced hearing, reduced color vision, and warm, **10**_____-covered bodies.

Scientists Discover a Distant Relative of Today's Horses (16)

Directions: Fill in a word that best completes each sentence.

An international team of researchers has discovered a genus of extinct horses that roamed North America during the last ice age. The new findings, published in the journal *eLife*, are based on an analysis of ancient DNA from fossils of the New World Stilt-Legged Horse excavated from North America.

Prior to this study, these thin-limbed, lightly built **1**_____ were thought to be related to the onager (which resembles a donkey), or simply a separate species within the genus *Equus*. The new results, however, reveal that these horses were not closely related to any living population of horses. "The evolutionary distance between the extinct stilt-legged horses and all living horses took us by surprise, but it presented us with an exciting opportunity to name a new **2**_____ of horse," said senior author Beth Shapiro.

Now named *Haringtonhippus francisci*, this extinct species of North American horse appears to have diverged from the main trunk of the family tree leading to Equus some 4 to 6 million years ago. The team named the new horse after Richard Harington, emeritus curator of Quaternary Paleontology at the Canadian Museum of Nature in Ottawa. Harington, who was not involved in the study, spent his career studying the ice age fossils of Canada's North and first described the stilt-**3**_____ horses in the 1970s.

There is a rich and deep fossil record of hooved mammals. Therefore, it has been a model system for understanding and teaching evolution. Two groups of living mammals are subclassified into orders by the number of toes that have. One group possesses an even number of toes with hooves (which are just enlarged toenails). This group includes pigs, hippopotamuses, camels, deer, and giraffes. The other group has an odd number of **4**_____ with hooves. It includes horses, ponies, donkeys, and zebras. Horses are members of the equine (equidae) family. This family consists of single hooved grazing animals. Now, in addition to the deep **5**_____ record, scientists can obtain ancient DNA, which often changes their understandings of how members of the **6**_____ family are related.

The new findings show that *Haringtonhippus* **7**_____ was a widespread and successful species throughout much of North America, living alongside populations of Equus but not interbreeding with them. In Canada's North, *Haringtonhippus* survived until roughly 17,000 years ago, more than 19,000 years later than previously known from this region. Although Equus survived in Eurasia after the last ice age, eventually leading to domestic horses, the **8**_____-legged *Haringtonhippus* was an evolutionary dead end.

Coauthor Eric Scott, a paleontologist at California State University San Bernardino, said that the **9**_____ of *Haringtonhippus* don't look much different from those of Equus. "But the DNA tells a fascinatingly different story altogether," he said. "That's what is so impressive about these findings. It took getting down to the molecular level to discern this new **10**_____."

Amber Discovery Indicates Lyme Disease is Older Than Human Race (17)

Directions: Fill in a word that best completes each sentence.

Lyme disease is frequently a misdiagnosed disease that was only recognized about 40 years ago, but new discoveries of ticks fossilized in amber show that *Borrelia burgdorferi*, the bacterium which causes Lyme has been around for 15 million years -- long before any humans walked on Earth. The researchers from Oregon State University studied 15-20 million-year-old amber from the Dominican Republic. This is the oldest fossil evidence ever found of *Borrelia* **1**_____.

"Ticks and the bacteria they carry are very opportunistic," said George Poinar, Jr., a professor emeritus in the Department of Integrative Biology of the OSU College of Science, and one of the world's leading experts on plant and animal life forms found preserved in amber. "In the United States, Europe and Asia, ticks are a more important insect vector of disease than mosquitos," Poinar said. "They can carry bacteria that cause a wide range of **2**_____, affect many different animal species, and often are not even understood or recognized by doctors.

"It's likely that many ailments in human history for which doctors had no explanation have been caused by tick-borne disease." Lyme disease is a perfect example. It can cause problems with joints, the heart and central nervous system, but researchers didn't even know it existed until 1975. If recognized early and treated with antibiotics, it can be cured. But it's often mistaken for other health conditions. And surging deer populations in many areas are causing a rapid increase in Lyme **3**_____.

The new research shows these problems with tick-**4**_____ disease have been around for millions of years. Bacteria are an ancient group that date back about 3.6 billion years, almost as old as the planet itself. As soft-bodied organisms, they are rarely preserved in the fossil record, but an exception is amber. Amber begins as a free-flowing tree sap, which hardens into a semi-precious mineral. As it does so, it traps and preserves soft-**5**_____ living organisms in death with the same detail they possessed in **6**_____.

A series of four ticks from Dominican amber were analyzed in this study, revealing a large population of cells that most closely resemble those of the present-day **7**_____ *burgdorferi* species. In 30 years of studying diseases revealed in the fossil record, Poinar has documented the ancient presence of such diseases as malaria, leishmania, and others.

Humans have probably been getting diseases, including Lyme disease, from **8**_____-borne bacteria as long as there have been humans, Poinar said. The oldest documented case is the Tyrolean iceman, a 5,300-year-old mummy found in a glacier in the Italian Alps. Before he was frozen in the glacier, the Tyrolean **9**_____ was probably already in misery from **10**_____ disease. Poinar said, "He had a lot of health problems and was really a mess."

Sixteenth Century Turkey Bones Found Under a Street in England (18)

Directions: Fill in a word that best completes each sentence.

Two femurs (thigh bones) and an ulna (wing) -- have been analyzed by University of Exeter archaeologists and identified as among some the first turkeys from the 16th century to be brought to England from the Americas. The bones are on display at the Royal Albert Memorial Museum & Art Gallery (RAMM) where Spanish, German and Italian pottery and glassware from the same site are also displayed. These items could have been on the table when the turkey dinner was served.

The first turkeys were introduced to England in 1524 or 1526 by William Strickland, a member of Parliament in the reign of Elizabeth the first, following a voyage to the Americas. Strickland is recorded to have bought six turkeys from Native American traders. When **1**_____ first appeared in **2**_____ they would have been a rare sight and the first ones are more likely to have been kept as pets for display of wealth rather than served as food. The bird became very popular after 1550 and already a common sight at Christmas dinners by the 1570s, before Thanksgiving in America was even invented.

The bones were found in 1983 as part of excavations at Paul Street, in central Exeter, before the building of a shopping center, but have never been identified or dated. Archaeologists at the University of **3**_____ have now examined the **4**_____ and, judging from pottery lying beside them, they date from the period 1520 to 1550. Professor Alan Outram, Head of Archaeology at Exeter, said: "This is an important discovery and could allow more research to be carried out about early domestic breeds and how the turkey has changed genetically since the 16th century."

Analysis by Malene Lauritsen, a post-graduate researcher in the **5**_____ of Exeter's **6**_____ department, has proved from the bones that the **7**_____ were butchered and were probably eaten as part of a feast by wealthy people. The Italian **8**_____ lying alongside was also of high quality. They were found together with the remains of a veal calf, several chickens, at least one goose and a sheep. This selection of food -- some of which were very expensive at the time -- suggests this was the rubbish created by a feast attended by **9**_____ people of high status.

The **10**_____ were found during excavations at Paul street, in central Exeter, and are part of the collections at Royal Albert Memorial Museum. Rachel Sutton, Lead Councilor for Economy & Culture and Deputy Leader of Exeter City Council said: "Exeter is blessed having a museum and a university that are both world-class. Working together, they are uncovering information about Exeter's past that would have been inconceivable only decades ago. Collaborations such as these are vital to Exeter's success."

Venomous Slow Loris May Have Evolved to Mimic Cobras (19)

Directions: Fill in a word that best completes each sentence.

Lorises are primates, relatives of monkeys, apes, and humans. They are nocturnal and found in tropical and woodland forests of India, Sri Lanka, and parts of southeast Asia. The Lorisinae family includes the genera slender lorises and slow lorises, of which there are several varieties. The **1**_____ lorises have more elongated bodies. Loris locomotion is a slow and cautious climbing form of quadrupedalism (walking on four limbs). Some lorises are almost entirely insectivorous, while others also include fruits, gums, leaves, and slugs in their diet.

Female lorises practice infant parking, leaving their young behind in nests. Before they do this, they bathe their **2**_____ with allergenic saliva that is acquired by licking patches on the insides of their elbows, which produce a mild toxin that discourages most predators, though orangutans occasionally eat lorises.

The slow loris has big doe eyes, a cute furry face, and tiny grasping hands, which are a deceptive mask for its deadly nature. Lorises are the only known venomous primate, secreting toxins from a gland located along the crook of their inner arms. When threatened, a loris will hiss and retreat into a defensive posture with its paws clasped on top of its head. This posture, of course, exposes the crook of their inner **3**_____. In this position, the loris's upraised arms combined with dark markings on its face look remarkably like the expanded hood of an angered Spectacled Cobra.

To add to the effect, slow lorises can even undulate like a Spectacled **4**_____. This unusual movement is made possible by an extra vertebra in their spines. The **5**_____ posture also allows slow lorises to suck the venom from their armpits and strike. The bites of these tiny primates have caused anaphylactic shock (a life-threatening allergic reaction) and even death in humans.

For eight million years, slow moving lorises and cobras have coexisted in the same parts of Asia. Together, they weathered through drastic climate changes, which may have forced the traditionally tropical forest dwelling **6**_____ to adapt to an open savannah-like environment by mimicking venomous snakes. Slow Lorises likely adopted **7**_____-like markings and movements as defense mechanisms. They are currently endangered animals.

Despite all its cunning evolutionary traits, the slow loris is endangered partly because of the exotic pet trade. "Knowledge of loris venom and its danger to **8**_____, we hope, will help curtail the growing illegal **9**_____ trade," Anna Nekaris, the author of the article published in the Journal of Venomous Animals and Toxins, and a professor of anthropology and primate conservation at Oxford Brookes University, says. Nekaris hopes more research will show people the loris is a rare and complex creature, not a cute **10**_____ face to domesticate.

Genetics

Abominable Snowman? Nope: Study Ties DNA Samples from Purported Yetis to Asian Bears (1)

Directions: Fill in a word that best completes each sentence.

The Tibetan Plateau in Asia is the highest plateau in the world. It is partly surrounded by the Himalayas and many of Earth's highest mountains. The Tibetan **1**_____ has a diverse habitat and biome with rich biological diversity. The colonization and population history of many species in the Tibetan Plateau remains poorly understood, despite current and future impacts of climate change and threats to diversity loss.

The Yeti or Abominable Snowman -- a mysterious, ape-like creature said to live in the mountains of **2**_____ – is a famous myth of Nepal and Tibet. Sightings have been reported for centuries. Footprints have been spotted. Stories have been passed from generation to generation. Although the human-like creature was known by several names in the Tibetan Plateau–Himalaya region, the species identity remained a mystery. There is a lack of conclusive evidence that the 'yeti' might be a known species.

A new DNA study has provided greater understandings about this **3**_____. The research, published in *Proceedings of the Royal Society B*, analyzed nine "Yeti" specimens, including bone, tooth, skin, hair and fecal samples collected in the Himalayas and **4**_____ Plateau. Of those, one turned out to be from a dog. The other eight were either from Asian black bears or Himalayan or Tibetan brown bears.

"Our findings strongly suggest that the biological underpinnings of the **5**_____ legend can be found in local bears, and our study demonstrates that genetics should be able to unravel other, similar mysteries," says lead scientist Charlotte Lindqvist, a visiting associate professor at Nanyang Technological University, Singapore (NTU Singapore).

Lindqvist says science can be useful in exploring the roots of **6**_____ about large, mysterious creatures. She notes that in Africa, the Western legend of an "African unicorn" was explained in the 20th century by British researchers who found an actual animal called the okapi, a giraffe relative that looks like a mix of giraffe, zebra, and horse. Similarly, in Australia the Aboriginal "Dreamtime" mythology probably draws from ancient encounters with large animals, which we can see from Australia's fossil record.

Besides tracing the origins of the Yeti **7**_____, Lindqvist's work is uncovering information about the evolutionary history of Asian bears. Bears in **8**_____ are either vulnerable or critically endangered from a conservation perspective, but not much is known about their history. Lindqvist explains that "The Himalayan brown **9**_____ are highly endangered. Clarifying population structure and genetic diversity can help in estimating population sizes and crafting management strategies."

"Further genetic **10**_____ on these rare and elusive animals may help illuminate the environmental history of the region, as well as bear evolutionary history worldwide -- and additional 'Yeti' samples could contribute to this work," Lindqvist says.

Why Some Cats Look Like They Are Wearing Tuxedos (2)

Directions: Fill in a word that best completes each sentence.

The cartoon character Sylvester is distinctive due to his black and white tuxedo-style coat. Cats with skin and fur marked by white patches in this way are known as bicolor or piebald. Piebaldism is also common in dogs, cows and pigs, deer, horses, and it appears more rarely in humans. It is caused by a mutation in a gene called KIT. **1**_____ usually manifests as white areas of fur, hair or skin due to the absence of pigment-producing cells in those regions. These areas usually arise on the front of an animal, often the belly and forehead. Piebald patterns are among the most striking animal coat patterns in nature.

Although the effects of piebaldism are relatively mild, it is one of a range of more serious defects called neurocristopathies. These result from defects in the development of tissues and can show up as heart problems, deafness, digestive problems and even cancer. The diseases are all linked by their reliance on a family of embryonic cells called neural crest cells. By understanding piebaldism better, scientists can improve our understanding of these related and more serious **2**_____.

Animals acquire piebald pigmentation patterns on their skin when they are still developing embryos. Current scientific understanding is that darkly colored pigment cells, which are thought to spread from the back to the front of the **3**_____, don't make it as far as the belly in time to pigment the hair and skin. This results in distinctive white patches of **4**_____ and **5**_____, usually around the belly of the animal. Normally, pigment cells starting near the back of the embryo multiply as they spread toward the **6**_____, which ensures all the skin is pigmented.

Researchers from the universities of Bath, Edinburgh and Oxford have found evidence that contradicts the idea of back-to-front cell migration. These findings show that cells in piebald animals migrate faster than in normal **7**_____, but that they don't divide as often. This means that there simply aren't enough cells to **8**_____ all the areas of the developing embryo. This may be true in chimeric animals (those resulting from the merging of two or more embryos/fertilized eggs) also, such as tortoiseshell cats. If the original embryos would have been differently colored (for example, black and white), the chimeric animal often has a patchy coat pattern. The predominant theory was that each patch was created by a small number of initiator cells that spread from back to front, but current research suggests that the pattern may simply be the result of several groups of cells of the same color coming together by chance.

This study used a combination of biological experimentation and complex mathematical modelling to demonstrate that pigment **9**_____ migrate randomly, rather than moving in a specific direction. Using this mathematical **10**_____, scientists can evaluate a range of possible hypotheses for pattern formation. They can potentially use the same model to investigate early development of certain types of nervous system cancer and other debilitating diseases.

First Primate Clones Produced Using the "Dolly" Method (3)

Directions: Fill in a word that best completes each sentence.

Scientists at the Chinese Academy of Sciences' Institute of Neuroscience, in Shanghai, published a report on two primate clones they produced using a technique called somatic cell nuclear transfer (SCNT). This advance brings scientists closer to a future in which they could create large numbers of genetically identical monkeys to serve as models for human diseases and other conditions. This could help researchers unravel complex questions, including how environmental factors may contribute to common human diseases.

There are ethical considerations that could limit this research, however. Many countries, including the US, have strict guidelines on primate research because of our close genetic relationship. Government biomedical research on chimpanzees in the US is effectively over, and all lab **1**_____ are being slowly retired. However, having a population of nonhuman **2**_____ that are genetically identical is a very powerful model for studying human **3**_____. This type of **4**_____ would need to be done on a case-by-case basis to determine if there are proper justifications for the procedures.

The technique used by the Chinese scientists, SCNT, has previously been used on more than 20 other species, including pigs and dogs. The numerous studies of animals produced by this method indicate they are as healthy as their non-cloned cousins.

The SCNT technique involves putting the nucleus of a donor cell into a fertilized egg that has had its own chromosomes removed. The **5**_____ then contains an exact copy of the donor's genome. After implantation and development in a surrogate mother, the eventual offspring will be a clone.

Up until the success of these Chinese scientists, the **6**_____ technique had not been achieved in primates. When researchers previously tried using the "Dolly" approach on monkeys it produced fetuses—but no pregnancy that lasted beyond 80 days. The main obstacle, the Institute of Neuroscience researchers wrote, was likely that transferred nuclei were not properly programmed to support embryonic development. This time, the Chinese team used two enzymes that removed genes' memory of being somatic cells (cells that make up tissues and organs). This extra step allowed development to proceed.

Although the success rate was low, scientists may be able to adjust their methods so that they could create dozens of clones soon. Researchers have cloned nonhuman primates in the **7**_____ using other techniques but using the SCNT **8**_____ is a major advance because it would likely be easier to use and reproduce **9**_____ numbers of **10**_____.

Aspects of monkey SCNT will also need to be improved before the technology can be used to produce primates for research. Although the advances are significant, from a practical use perspective, the pregnancy and live birth rate were not high enough for the technique to be used on a wide scale.

Cold Snap Shapes Lizard Survivors (4)

Directions: Fill in a word that best completes each sentence.

In January 2014, an epic cold wave swept across the southeast—a "snowpocalypse" so severe that thousands of drivers in Atlanta abandoned their **1**_____ on icy highways and Interstates.

Shane Campbell-Staton was watching it all unfold from Harvard, where he was getting his PhD. He'd just wrapped up his last field season in Texas, studying the green anole lizard. And as he was scanning photos of the storm, he came across something unexpected: a photo that included his research subject. "There was this one picture of a green anole that was upside down, dead in the snow. And it was sort of a Eureka moment. And I thought to myself, well maybe I should go back out and see if these populations I'd been studying, if they showed any sort of response to this pretty extreme **2**_____ event in the south."

Campbell-Staton had been studying the cold tolerance of different populations of these green anole **3**_____ and the cold snap had just delivered the perfect experiment—a chance to see natural selection in action. He went to Atlanta in April right after the **4**_____ storms had subsided. He noticed that in the southernmost population the survivors of the storm could maintain function at significantly colder temperatures than the population before the storm. This ability to maintain function at **5**_____ temperatures is something that is typically seen much farther north.

He did genetic analyses and found that the genes switched on in the surviving southern lizards overlapped with **6**_____ more typically turned on in their cold-hardy northern cousins. The southern survivors also carried variations in their DNA that more closely matched **7**_____ lizards. The cold tolerance, gene expression, and gene variants the southern lizards carried suggested the winter storm had indeed caused selection on the southern lizards.

Campbell-Staton, now at the University of Illinois and the University of Montana, is quick to mention that this isn't quite evolution yet. It's just one generation—he hasn't yet seen these traits passed down to another set of **8**_____. That's his next investigation.

It is unknown if this change in the population is a good thing. If another cold wave rolls through, the surviving southern population will be better prepared—more **9**_____-tolerant. But selection comes at a cost: death. The lizards that **10**_____ during this snow event may have had genetic variants that would have allowed them to survive a heat wave, or a drought, or some other extreme event and now those genes are gone.

Doctors Fix Genetic Defect and Grow Boy New Skin (5)

Directions: Fill in a word that best completes each sentence.

Doctors in Germany have used gene therapy to grow a complete new epidermis for a boy with a life-threatening inherited skin disease. The **1**_____ was seven years old when he was admitted to the hospital in 2015 with severe complications of junctional epidemolysis bullosa (JEB). This is a rare skin condition caused by faults in a few genes that code for anchor molecules that glue the surface skin layer, called the epidermis, to the deeper skin layer known as the dermis.

Patients with JEB have extremely fragile skin. Even very mild trauma or minor abrasions lead to severe and painful blistering, ulceration and scarring. The break-down in the skin also makes patients more likely to get bacterial infections. Because of constant demands placed on skin stem cells, skin cancer is a complication.

When the boy came to the hospital, he'd already lost over half of the skin from his body, and he was suffering overwhelming **2**_____ infections. Faced with these extreme circumstances, doctors resorted to an experimental treatment - they used gene therapy to fix the genetic fault responsible for his **3**_____ in some **4**_____ cells, and then used tissue culture to grow a replacement epidermis.

The new skin was prepared from a patch of healthy skin taken from the boy's groin area. Skin **5**_____ cells called keratinocytes were taken from this 4 cm^2 piece of tissue and infected with a harmless virus, which had been reprogrammed to carry a healthy copy of the LAMB3 gene that was defective in the patient. The **6**_____ inserted the healthy LAMB3 gene into the DNA of the skin stem cells, restoring their ability to make healthy amounts of the protein Laminin-5, which gives skin its strength.

The boy's corrected stem cells were grown in culture, either on a plastic scaffolding or on a matrix of a molecule called fibrin, which the body naturally uses to make blood clots and provides a strong foundation for tissue to grow.

Over about 12 weeks, the German team were able to **7**_____ nearly a square meter of new skin from the genetically repaired stem cells. Sheets of this new **8**_____ were transplanted onto the patient, replacing his damaged tissue and meshed together like a patchwork. By the end of his treatment, nearly all the patient's **9**_____ was covered with the treated skin.

Now, two years later, the patient is well and back at school. His skin is healthy. He does not need to use any medication, and if he cuts himself, his **10**_____ heals normally. The team, who have published their remarkable work in the journal *Nature*, will continue to monitor him closely because there is a chance that genetic changes introduced into the repaired stem cells during the procedure could lead to cancer. At this point, however, tests done at the time of transplant and since then have not shown any evidence of negative effects.

You May Be as Friendly as Your Genes (6)

Directions: Fill in a word that best completes each sentence.

Why some individuals seek social engagement and friendship while others shy away may be dependent on the expression and sequence of two **1**_____ in their bodies. A group of researchers from the National University of Singapore (NUS) has found that adults who have higher expression of the CD38 gene as well as differences in CD157 gene sequence are more socially adept than others. CD38 and CD157 genes regulate the release of oxytocin, the paramount social hormone in humans involved in behaviors such as pair-bonding, mating, and child-rearing, and more sophisticated behaviors such as empathy, trust and generosity.

This study of gene expression (i.e. how much of a gene is produced in the body) demonstrates the importance of the oxytocin network on shaping **2**_____ and communication skills that are instrumental in building friendships. The findings were published in the scientific journal *Psychoneuroendocrinology*.

The research team from the NUS studied 1,300 Chinese adults in Singapore in a non-clinical setting. They investigated the correlation between the expression of the CD38 gene and CD157 gene sequence, both of which have been implicated in autism studies, and an individual's social skills as captured by three different questionnaires. These questionnaires evaluated participants' ability to engage in relationships; their value on the importance of friendships as well as the number of close **3**_____ they have.

The results from the study showed that **4**_____ with higher expression of CD38 have more close **5**_____. This association was observed more prevalently among the male participants. Participants with lower CD38 expression reported fewer social skills such as difficulty in "reading between the lines" or engaging less in social chitchat and tending to have fewer friends.

The researchers found that oxytocin, and the CD38 and CD157 **6**_____ that govern its release, contribute to differences in social skills from one extreme of having many good peer **7**_____ to the other extreme of avoiding contacts with other people. This far extreme is one of the characteristics of autism. There is no cause for worry, however. Most people fall between the two **8**_____.

While expressed genes can influence behaviors, experiences can influence the expression of genes in return. Social environments can impact whether the genes are **9**_____. For most people, having supportive families, friends, and colleagues would lessen the effects from disadvantageous genes.

New drugs could be developed that mimic or enhance the functions of the CD38 and CD157 genes. If proven viable, future therapies may help those clinically determined to have difficulty maintaining social and working **10**_____ with others so that they could live a better quality of life.

Researchers Find Genetic 'Dial' Can Control Body Size in Pigs (7)

Directions: Fill in a word that best completes each sentence.

Researchers from North Carolina State University have demonstrated a connection between the expression of the HMGA2 gene and body size in pigs. The work further demonstrates the gene's importance in body **1**_____ regulation across mammalian species and provides a target for gene modification.

"Essentially, HMGA2 is a gene that controls the total number of cells that an animal has," says Jorge Piedrahita, Director of the Comparative Medicine Institute at NC State. "The **2**_____ is only active during fetal development, and 'programs' in the number of cells that the animal will be able to generate. When the animal is born, it will only be able to grow to the size dictated by the number of **3**_____ that it can produce."

Researchers had previously studied the HMGA2 analogue (genetic comparison) in mice, which have two different genes (HMGA2 and HMGA1) involved in **4**_____ size and body mass index determination. In **5**_____, inactivation of one alleles of HMGA2 results in a body size reduction of 20%; inactivation of both alleles of HMGA2 results in a body size reduction of 60%. Likewise, in humans, micro-deletions involving the HMGA2 locus result in short stature, suggesting the function of the HMGA2 protein is found among all mammals.

Since pigs and humans share the **6**_____ gene responsible for growth regulation in their species, the NC State study looked at body size in pigs that expressed both copies of the gene, one copy, or neither copy. "We found that the amount of the gene expressed is proportional to the size of the animal," Piedrahita says. "If both copies were expressed the pig was 'normal' sized. If one copy was expressed the **7**_____ was roughly 25 percent smaller than normal, and if neither **8**_____ was expressed the pig was 75 percent smaller."

In cases where both copies of the gene were deleted, the pigs did grow and develop, but they were sterile. "Overall, it seems that controlling the expression of HMGA2 is like using a dial to control body size," said Piedrahita.

Other findings of the research study were that the deletion of HMGA2 affected the resources that the pig fetuses received in utero. In litters containing fetuses with both copies of the gene deleted and fetuses with one or more copy of the gene expressed, the **9**_____ with both copies deleted did not survive the pregnancy. However, if the litter only contained fetuses with both copies deleted, the fetuses survived and developed normally.

Overall, the study shows that the effect of HMGA2 with respect to growth regulation is highly conserved among mammals. This finding opens the possibility of regulating body and organ size in a variety of mammalian **10**_____ including food and companion animals.

Can Gene Editing Save the World's Chocolate? (8)

Directions: Fill in a word that best completes each sentence.

Scientists are racing to save the cacao tree because fungi and viruses are infecting them. **1**_____ trees (*Theobroma cacao*) grow in tropical environments, within about 20 degrees north and 20 degrees south of the equator. These trees sprout colorful, football-size pods containing beans used to make chocolate. Unfortunately for chocolate lovers, fungi also flourish in **2**_____ environments and can easily infect entire cacao tree farms, causing harmful conditions such as frosty pod, black pod and witch's broom, according to a 2016 report from the National Oceanic and Atmospheric Administration.

"Cacao can be afflicted by several devastating conditions," said Brian Staskawicz, a professor in the Department of Plant and Microbial Biology at the University of California, Berkeley. "We're developing CRISPR editing technologies to alter the DNA in cacao plants to become more resistant to both viral and fungal diseases."

Human-caused climate change is also putting the trees at risk, as rising temperatures caused by greenhouse-gas emissions may alter climatic conditions where cacao **3**_____ typically grow, mainly in West Africa and Indonesia. As the mercury rises and squeezes more water out of soil and plants, scientists believe it is unlikely that rainfall will increase enough to offset the moisture loss. That means cacao production areas are set to be pushed thousands of feet uphill into mountainous terrain which is carefully preserved for wildlife. Officials in countries such as Côte d'Ivoire and Ghana - which produce more than half of the world's chocolate - will face a difficult decision: whether to maintain the world's supply of chocolate or to save their dying ecosystems.

4_____ is more than a delicious treat. Cacao helps employ up to 50 million people worldwide, according to the World Cocoa Foundation. In an effort to save the tree and its crop, Mars Inc.—which makes M&M's, 3 Musketeers and Snickers—has teamed up with scientists at the Innovative Genomics Institute (IGI) to engineer trees that are resistant to certain fungi and **5**_____.

Their main tool is CRISPR-Cas9, a pair of molecular scissors that can precisely cut out chunks of DNA and replace them with new stretches of **6**_____. Myeong-Je Cho, director of plant genomics and transformation at IGI, is already working with cacao seedlings, looking for ways to help cacao growers stay put even as the climate warms and fungi invade their farms.

CRISPR-Cas9 will likely help researchers find **7**_____- and virus-resistant trees sooner than cross pollinating plants the old-fashioned way. **8**_____ trees take between five and seven years to grow their colorful pods, and it isn't clear whether these **9**_____ will be susceptible to disease until they are grown. With **10**_____-Cas9, scientists can engineer the plants to be resistant from their most immature stages.

Gene Therapy Restores Sight to Blind Mice (9)

Directions: Fill in a word that best completes each sentence.

The retina is the light-sensitive tissue at the back of the eye where rod and cone cells convert light waves into electrical signals that the brain can understand. In diseases like retinitis pigmentosa, the rod and cone cells of the **1**_____ die off, leaving patients unable to see. Samantha De Silva and her colleagues have developed a way to use a harmless virus to deliver genetic instructions that can make other healthy cells in the retina become light sensitive and take over the role of the missing **2**_____ and cones.

De Silva and her colleagues created a harmless virus, called AAV, that has a copy of the human melanopsin gene, known as OPN4. This codes for a **3**_____ sensitive molecule that is used by one group of cells in the retina to detect light at the blue end of the spectrum. This research is published in the journal PNAS.

The virus with the melanopsin gene was injected beneath the retina in a group of mice with a form of blindness like retinitis pigmentosa. Alongside a control group of un-injected animals, the mice were tested after the treatment to assess their vision and to see whether their pupils responded to light, whether the retina generated the correct pattern of electrical signals when light fell upon it, and if the animals could recognize objects, indicating that retinal signals were being sent correctly to their brains.

The tests showed that over 40% of retinal cells with the **4**_____ gene fired off impulses when light shone on them compared with only 18% in the control, untreated mice. Pulses of light shone into the eyes also caused blood flow changes in the visual areas of the animals' brains, which may reflect neurological processing of the signals arriving from the eye. Pupil constriction in response to light was also present, and the treated mice spent far less time in brightly lit areas compared with the control animals. The virus-injected mice also showed improved ability to recognize objects in their surroundings, suggesting that they could make sense of the visual information being presented to their nervous systems.

Examination of the retinae from the mice shows that the viral injections with the melanopsin **5**_____ had accessed a range of cells in the retina that are not normally involved in detecting light directly. Adding the melanopsin gene to these cells had made them light **6**_____ so that they could partly take over the function of the missing rods and **7**_____. Unfortunately, the melanopsin **8**_____ in these new cells does not act as fast as the molecule it was replacing. Therefore, it's less helpful for interpreting movement than seeing static objects, like the location of a door. Still, it's a step forward!

More good news is that because melanopsin is naturally present in the human retina, it's less likely to trigger an immune response when more is added in **9**_____ therapy. Human clinical trials are expected in 2020. "Fortunately, there are other retinal gene **10**_____ clinical trials on-going in Oxford, so we can speed up the route to the clinic for this new approach," says De Silva.

Gray Hair Linked to Immune System Activity and Viral Infection (10)

Directions: Fill in a word that best completes each sentence.

A new study on mice offers insights into why some people's hair may turn gray in response to a serious illness or chronic stress. Published in the open access journal *PLOS Biology*, researchers at the National Institutes of Health and the University of Alabama, Birmingham have discovered a connection between the genes that contribute to hair color and the genes that notify our bodies of a pathogenic infection.

When a body is under attack from a virus or bacteria, the innate immune system kicks into gear. All cells can detect foreign invaders and they respond by producing signaling molecules called interferons. Interferons signal to other **1**_____ to act by turning on the expression of genes that inhibit viral replication, activate immune effector cells, and increase host defenses.

Melissa Harris, of the Department of Biology at UAB explains, "Genomic tools allow us to assess how all of the **2**_____ within our genome change their expression under different conditions, and sometimes they change in ways that we don't anticipate. We are interested in genes that affect how our stem cells are maintained over time. We like to study gray hair because it's an easy read-out of melanocyte stem cell dysfunction." Melanocyte stem cells are essential to hair color as they produce the melanocytes that are responsible for making and depositing pigment into the hair shaft.

In this case, an unexpected link was found between **3**_____ hair, the transcription factor MITF, and innate immunity. MITF is best known for its role in regulating the many functions within melanocytes. But the researchers found that **4**_____ also serves to keep the melanocytes' interferon response in check. If MITF's control of the **5**_____ response is lost in melanocyte stem cells, hair-graying results. Furthermore, if innate immune signaling is artificially activated in mice that are predisposed for getting gray **6**_____, increased numbers of gray hairs are also produced.

"This new discovery suggests that genes that control pigment in hair and skin also work to control the innate **7**_____ system," said William Pavan, of the Genetic Disease Research Branch at NIH's National Human Genome Research Institute (NHGRI). "These results may enhance our understanding of hair graying. More importantly, discovering this connection will help us understand pigmentation diseases with **8**_____ immune system involvement like vitiligo." **9**_____, which causes discolored skin patches, affects between 0.5 percent to 1 percent of all humans.

Why mice that are predisposed for getting gray hair are more susceptible to dysregulated innate immune signaling remains to be answered. The researchers speculate that perhaps this can explain why some people experience premature gray hair early in **10**_____, and they will continue their studies to address this question.

The Happy Hour Gene (11)

Directions: Fill in a word that best completes each sentence.

People who are very sensitive to alcohol tend to drink less. People who can drink everybody under the table are more likely to become alcoholics. Researchers studying the problem of alcoholism used fruit flies as research subjects. Laboratory fruit **1**_____ aren't so different from people when they drink. They go through a phase of hyperactivity, they gradually become uncoordinated, stop moving, and eventually fall over.

Just like people, however, some **2**_____ flies are more tolerant than others of alcohol. Those alcohol-tolerant fruit flies possessed a version of a gene, aptly named the Happy Hour Gene, which does not inhibit (stop) the effects of a protein called epidermal growth factor, or EGF. The normal version of the Happy **3**_____ Gene does inhibit EGF.

A cancer drug, Tarceva, inhibits the cellular proteins stimulated by EGF. Researchers gave Tarceva to rats that were used to consuming large amounts of alcohol. They then presented the rats with both alcohol and water, and let the animals make the choice. The rats reduced their alcohol consumption, favoring **4**_____.

Ulrike Heberlein, the molecular biologist at the University of California, San Francisco, who led the study, would like to find out whether **5**_____ has the same effect in humans. She notes that it took relatively low doses of Tarceva in rats to reduce **6**_____ consumption, so there's some hope that it would also be possible to use a low dose of the drug in people. That's important, since it could be a problem to give high doses of powerful **7**_____ drugs to patients who don't have cancer.

Happy Hour isn't the only fruit fly gene associated with alcohol metabolism. Heberlein's lab also discovered other fruit fly genes associated with alcohol metabolism. Among them: Cheapdate, a gene mutation causing super-sensitivity to alcohol and a gene called Hangover, which helps builds alcohol tolerance over time.

Heberlein's research into alcohol **8**_____ was funded by the National Institute on Alcohol Abuse and Alcoholism, the Department of Defense and the state of California. **9**_____ Hour is one out of many genes, plus environmental factors, that likely contribute to alcoholism. The "ultimate drug" to fight **10**_____ might be a cocktail that targets a bunch of things at once, Heberlein said.

Is "Junk DNA" What Makes Humans Unique? (12)

Directions: Fill in a word that best completes each sentence.

Even though chimpanzees appear very different from humans, our two species are strikingly similar on the genetic level. The parts of our DNA that contain instructions for making proteins—the building blocks of our bodies—differ by less than 1 percent. Interestingly, genes that code for **1**_____ are only a small part of our genomes. Some of the biggest differences between humans and chimps are in the parts of DNA that are outside of the genes, the area scientists once called "junk DNA".

Katherine Pollard of the University of California identified areas of these "**2**_____ DNA" regions that have changed quickly between humans and chips. They have been named human accelerated regions (HAR).

Pollard and Nadav Ahituv, a geneticist who runs a separate lab at U.C.S.F., were able to create a method for converting human and chimpanzee skin cells into pluripotent stem cells, which have the potential to become nearly any other cell type. Although these **3**_____ stem cells could have become any kind of cell, the team chose neurons because intelligence is the most distinctive human trait. Pollard and Ahituv created thousands of **4**_____ at a time and spliced the HAR DNA into those cells. Then they examined what the HARs did at two different points in the cells' development.

They found almost half of these pieces of DNA—which do not appear naturally in the chimpanzee genome—were active in the growing neurons. The HARs were not producing proteins; they were in the "junk **5**_____". The result surprised Ahituv: "This is the first comprehensive study of all these sequences, and it shows that 43 percent of them...could have a functional role in neural development."

According to Pollard, the parts of the chimpanzee **6**_____ that are analogous to the HARs have not changed at all in millions of years, and they are nearly identical to the same regions in most animals. Pollard says natural selection was acting to keep these parts of these animals' genomes from changing, but that is not true for humans. Most of [the HARs] in humans have many random mutations. To accelerate the process, the individuals carrying those changes produced more offspring. Scientists don't know what caused this to happen, but the fact that so many HARs are involved in neuronal development suggests the change may have had something to do with the evolution of **7**_____.

These changes, however, came with some severe downsides. "A lot of these **8**_____ lie near genes that are associated with human-specific disease like autism and schizophrenia," Ahituv says. The team found the individual mutations would increase or decrease the amount of protein a gene was producing. Essentially, natural selection was fine-tuning how the **9**_____ were expressed because too much or too little of a specific **10**_____ can cause problems. The findings in this study may be helpful to medical researchers by showing them what parts of the genome to target for new therapies for mental health diseases.

Link Between Biological Clock and Aging Revealed (13)

Directions: Fill in a word that best completes each sentence.

Scientists studying how aging affects the biological clock's control of metabolism have discovered that a low-calorie diet helps keep these energy-regulating processes running smoothly and the body younger. In a study in the journal, *Cell*, Paolo Sassone-Corsi, director of the Center for Epigenetics & Metabolism at the University of California, Irvine, reveals how circadian rhythms -- or the body's biological clock -- change because of normal aging. The clock-controlled circuit that directly connects to the process of aging is based on efficient metabolism of energy within **1**_____.

The Sassone-Corsi team tested the same group of mice at 6 months and 18 months, drawing tissue samples from the liver, the **2**_____ which operates as the interface between nutrition and energy distribution in the body. Energy is metabolized within cells under precise circadian controls.

The researchers found that there were notable changes in the circadian mechanism that turns genes on and off based upon the cellular **3**_____ usage in the mice at 18 months. Their older cells processed energy inefficiently. The circadian rhythm that turns genes on and **4**_____ based on energy usage works great in a **5**_____ animal, but it basically shuts off in an old mouse.

A second group of older mice that were fed a diet with 30 percent fewer calories for six months, energy processing within cells was improved. Caloric restriction works by rejuvenating the biological **6**_____.

In another study, a research team from the Barcelona Institute for Research in Biomedicine collaborated with the Sassone-Corsi team to test body clock functioning in stem cells from the skin of young and older mice. They also found that a low-calorie **7**_____ helped maintain most of the rhythmic functions of youth.

The low-calorie diet greatly contributes to preventing the effects of normal **8**_____. Keeping the rhythm of stem cells 'young' is important because stem cells serve to renew and preserve day-night cycles in tissue. Eating less appears to prevent tissue aging and, therefore, prevent stem cells from reprogramming their circadian activities. The implications of these studies for human aging could be far-reaching. Once the link that promotes or delays aging has been identified, treatments can be developed to regulate this link.

It's previously been shown in fruit fly studies that low-calorie diets can extend longevity, but these most recent studies are the first to show that **9**_____ restriction influences the body's **10**_____ rhythms' involvement with the aging process in cells.

New Way to Activate Stem Cells to Make Hair Grow (14)

Directions: Fill in a word that best completes each sentence.

UCLA researchers have discovered a new way to activate the stem cells in the hair follicle to make hair grow. The research, led by scientists Heather Christofk and William Lowry, may lead to new drugs that could promote hair **1**_____ for **2**_____ with baldness or alopecia, which is hair loss associated with such factors as hormonal imbalance, stress, aging or chemotherapy treatment. The research was published in the journal Nature Cell Biology.

Hair follicle stem cells are long-lived cells in the hair follicle; they are present in the skin and produce hair throughout a person's lifetime. They are "quiescent," meaning they are normally inactive, but they quickly **3**_____ during a new hair cycle, which is when new hair growth occurs. The quiescence of hair follicle stem cells is regulated by many factors. In certain cases, they fail to activate, which is what causes hair loss.

In this study, Christofk and Lowry, of Eli and Edythe Broad Center of Regenerative Medicine and Stem Cell Research at UCLA, found that hair follicle stem cell metabolism is different from other cells of the skin. Cellular metabolism involves the breakdown of the nutrients (including glucose) needed for cells to divide, make energy, and respond to their environment. The process of **4**_____ uses enzymes that alter these nutrients to produce "metabolites." As hair follicle stem cells consume the **5**_____ glucose from the bloodstream, they process the glucose to eventually produce a metabolite called pyruvate. The cells then can either send pyruvate to their mitochondria -- the part of the cell that creates energy -- or can convert **6**_____ into another metabolite called lactate.

The study team began to examine whether decreasing the entry of pyruvate into the mitochondria would force hair follicle **7**_____ cells to make more lactate, and if that would activate the cells and grow hair more quickly. First the production of lactate in mice was blocked and this prevented hair follicle stem cell activation. Later, lactate production in the mice was **8**_____.

The researchers observed that the increase in lactate production accelerated hair follicle stem cell activation, which increased the hair cycle! This new information allowed for the investigation of potential drugs that could combat hair **9**_____.

Two different drugs which influence hair follicle stem cells to promote **10**_____ production have been studied in mice. The idea of using drugs to stimulate hair growth through hair follicle stem cells is very promising given how many millions of people, both men and women, deal with hair loss. Currently, the two drugs have not been tested in humans or approved by the Food and Drug Administration as safe and effective for use in humans.

Tree Rings Used to Counter Smugglers' Rings (15)

Directions: Fill in a word that best completes each sentence.

Illegal logging is a serious problem. It costs billions of euros (EUR, the currency of the European Union) annually, endangers people's jobs, biodiversity, food security, as well as damaging the environment by increasing carbon dioxide (CO2) emissions. "By undermining sustainable forest management, **1**_____ loggers are damaging natural ecosystems," said Professor Pieter Zuidema, of the Netherlands. "If these forests degrade, they will release CO2 into the atmosphere and contribute to climate change."

Regulations have been in place since 2004 prohibiting illegally logged timber from entering the European market but identifying illegally traded wood is difficult. Currently, customs officers still rely on paper trails to identify imported timber. These documents can be forged by criminals or bought illegally from corrupt officials in the territories through which contraband **2**_____ is smuggled.

Authorities need global tracking systems that smugglers cannot tamper with. Prof. Zuidema is looking at a solution that uses the traceability that nature stamps deep inside the wood itself, its DNA. Prof. Zuidema has spent years studying the chemical composition of trees from across the tropics and has combed through the DNA of a tropical tree species called Tali. The **3**_____ grows in large areas of Africa. Like human populations, each region the trees grow in evolves its own familial traits. By measuring mutations in the tree **4**_____, Prof. Zuidema can trace back Tali timber to its geographic origins. These records make it possible to verify the degree to which the DNA of timber reaching the EU is related to the specimens he has sampled in the wild.

In a blind test, Prof. Zuidema's technique identified **5**_____ from Congo and Cameroon with an accuracy of 90%. In some cases, geneticists narrowed down the point of origin to under 15 kilometers. This is precise enough to distinguish between legal timber logged in licensed forests and **6**_____ timber stolen from neighboring nature reserves.

Elsewhere, scientists are working on a handheld electronic device that could trace timber by flashing tree rings with near-infrared beams and inferring the wood's species and origin from the reflected radiation. The researchers are first charting out the near-**7**_____ response of thousands of trees in different areas and then using the power of big data to home in on their point of origin. At present, **8**_____-infrared scanners cost upwards of EUR 40 000, but researchers are trying to bring prices closer to **9**_____ 5 000. If they do, the technology could prove useful not only to law enforcement, but also to EU companies engaged in wood-based manufacturing because they can be held accountable for any illegal wood products that they sell.

Researchers say the challenge is in building a reference database with tree samples from over 100 species and more than 30 countries around the world. When near-infrared records from up to 120,000 trees are uploaded, a quick check on the cloud would trace the **10**_____ of timber entering the EU.

Simple Urine Test Could Measure How Much Our Body Has Aged (16)

Directions: Fill in a word that best completes each sentence.

While everyone born in the same year has the same chronological age, the bodies of different people age at different rates. This means that, although the risk of many diseases increases with age, the link between our **1**_____ in years and our health and lifespan is relatively loose. Many people enjoy long lives, relatively free of disease, while others suffer chronic illness and premature death.

Some researchers consider normal aging to be a disease, where our cells accumulate damage over time. The rate of this cellular **2**_____ can vary from person to person, and may be dictated by genetics, lifestyle and the environment we live in. This cellular damage may be a more accurate indication of our biological age than the number of years since we were born.

Finding a way to measure **3**_____ age could help to predict the risk of developing age-related disease and even death. Being able to measure biological age is also important in order to determine if treatments developed to slow aging are effective.

One mechanism believed to cause biological aging involves a molecule vital to our survival: oxygen. "Oxygen by-products produced during normal metabolism can cause oxidative damage to biomolecules in cells, such as DNA and RNA," explains Jian-Ping Cai, author of a study on biological age, which was published in *Frontiers in Aging Neuroscience*. "As we age, we suffer increasing **4**_____ damage, and so the levels of oxidative markers increase in our body."

One oxidative **5**_____ is 8-oxoGsn. This marker results from oxidation of a crucial molecule in our cells called RNA. In previous studies in animals, Cai and colleagues found that 8-oxoGsn levels increase in urine with age. To see if this is true for humans as well, the researchers measured **6**_____ in urine samples from 1,228 Chinese residents aged 2-90 years old, using a rapid analysis technique called ultra-high-performance liquid chromatography. "We found an age-dependent increase in urinary 8-oxoGsn in participants 21 years old and older." said Cai. "Therefore, urinary 8-oxoGsn is promising as a new marker of aging."

Interestingly, levels of 8-oxoGsn were roughly the same between men and **7**_____, except in post-menopausal women, who showed higher levels. This may have been caused by the decrease in estrogen levels that happens during menopause, as **8**_____ is known to have anti-oxidant effects. The team's rapid analysis technique could be useful for large-scale aging studies, as it can process **9**_____ samples from up to 10 participants per hour. "Urinary 8-oxoGsn may reflect the real condition of our bodies better than our chronological age and may help us to predict the risk of age-related **10**_____," concludes Cai.

Where do heart cells come from? (17)

Directions: Fill in a word that best completes each sentence.

Researchers from Sanford Burnham Prebys Medical Discovery Institute (SBP), the Cardiovascular Institute at Stanford University, and several other institutions, were surprised to discover that the four genes in the Id family play a crucial role in heart development, telling undifferentiated stem cells to form heart tubes and eventually muscle. While Id genes have long been known for their activity in neurons and blood cells, this is the first time they've been linked to **1**_____ development. These findings give scientists a new tool to create large numbers of cardiac cells to regenerate damaged heart tissue. The study was published in the journal Genes & Development.

It has always been unclear what process begins cardiac cell development from **2**_____ cells. Now scientists know these four **3**_____ in the Id family are the earliest determinants of cardiac cell fate. Knowing this enables researchers to generate unlimited amounts of cardiac progenitors (parent cells) for regenerative purposes, disease modeling, and drug discovery.

The international team, led by Alexandre Colas, Ph.D., assistant professor in the Development, Aging and Regeneration Program at SBP, used several techniques to identify the role Id genes play in heart **4**_____.

The technique CRISPR played a crucial role, allowing scientists to knock out all four **5**_____ genes. Previous studies had knocked out some of these genes, which led to damaged hearts. However, removing all four genes created mouse embryos with no hearts at all. This discovery comes after a decades-long effort to **6**_____ the genes responsible for heart development.

This is a completely unanticipated pathway in making the heart. Scientists have been working for many years to figure out how the heart is specified during **7**_____. This technology could have wide-spread impact throughout biology.

The discovery of the role of these four **8**_____ in organ formation shows that researchers can effectively search through the human genome to find genes that control complex biology, like making **9**_____ cells or causing disease. Understanding this pathway could ultimately jumpstart efforts to use stem cells to generate heart **10**_____ and replace damaged tissue. In addition, because Id proteins are the earliest known mechanism to control cardiac cell fate, this work is an important milestone in understanding cardiovascular developmental biology.

Human Body Systems

Autism Symptoms Improve After Fecal Transplant, Small Study Finds (1)

Directions: Fill in a word that best completes each sentence.

Children with autism may benefit from fecal transplants -- a method of introducing donated healthy microbes into people with gastrointestinal disease to rebalance the gut, a new study has found. In the **1**_____, which appears in the journal *Microbiome*, 18 children with autism and moderate to severe gastrointestinal problems received **2**_____ transplants. The children included both males and **3**_____ 7 to 16 years old. Parents and doctors said they saw positive changes that lasted at least eight weeks after the treatment. Children without autism were included for comparison of bacterial and viral gut composition prior to the study.

"Transplants are working for people with other **4**_____ problems. And, with autism, gastrointestinal symptoms are often severe, so we thought this could be potentially valuable," said Ann Gregory, one of the study's lead authors.

Parents of the children not only reported a decrease in gut problems, including diarrhea and stomach pain in the eight weeks following the end of treatment, they also said they saw significant improvements in behavioral autism symptoms. The researchers collected this information from parents through established, standardized questionnaires to assess social skills, irritability, hyperactivity, communication and other measures. At the end of the study, the bacterial diversity in the **5**_____ with autism was indistinguishable from the children **6**_____ autism.

Fecal transplantation is done by processing donor feces and screening it for disease-causing viruses and bacteria before introducing it into another person's gastrointestinal tract. In this study, the researchers used a method called microbiota transfer therapy, which started with the children receiving a two-week course of antibiotics to wipe out much of their existing gut flora. Then, doctors gave them an initial high-dose fecal **7**_____ in liquid form. In the seven to eight weeks that followed, the children drank smoothies blended with a lower-dose powder.

There currently exists no approved pharmaceutical treatment for autism. James Adams, one of the study's lead authors and an Arizona State University professor who specializes in **8**_____, called the results compelling, but cautioned that more rigorous studies confirming benefits must be done before the approach could be used widely. Limitations of this study include its small sample **9**_____. The children and their parents also knew they were an experimental group (neither the researchers nor the subjects were blinded). There was no formal control **10**_____ of children with autism that did not receive a fecal transplant. Also, researchers relied on parents' observations. Both limitations decrease the validity of the experiment.

"We have to be mindful of the placebo effect and we have to take it with a grain of salt," said Matthew Sullivan, an associate professor of microbiology at Ohio State. "But it does give us hope." The research team is seeking additional funding for a larger clinical trial.

Many Babies Healthier in Homes with Dogs (2)

Directions: Fill in a word that best completes each sentence.

Babies in homes with dogs have fewer colds, fewer ear infections, and need fewer antibiotics in their first year of life than **1**_____ raised in pet-free homes. Homes with cats are healthier for babies, too, but not to the same extent as those with **2**_____, notes researcher Eija Bergroth, MD, of Finland's Kuopio University Hospital. "The strongest effect was seen with dog contacts. We do not know why it was stronger than with **3**_____ contacts. It might have something to do with dirt brought inside by the dogs, especially since the strongest protective effect was seen with children living in houses where dogs spent a lot of time outside," says Bergroth.

To take a closer look at the situation, Bergroth's team followed 397 Finnish children from their third trimester of pregnancy through their first 12 months of life. Parents filled out weekly diaries with detailed information on their child's health and on their child's contact with dogs and cats. "It might have something to do with the dog itself as an animal," Bergroth suggests. "The living environment can also affect this. These children lived in rural or suburban areas, so inner-city kids -- and dogs -- might get different results."

A time-honored theory, the hygiene hypothesis, suggests that children's immune systems mature best when infants are exposed to germs in just the right amount. Too many **4**_____ are unhealthy, but so is a sterile, germ-free home. That theory is now giving way to the "microbiome hypothesis," says Karen DeMuth, MD, MPH, assistant professor of pediatrics at Atlanta's Emory University. "The **5**_____ hypothesis is that early-life exposure to wide varieties of microbes lets them mix with the microbes in the gut and helps them keep the **6**_____ system from reacting against itself and causing autoimmune disease, or from reacting against stuff you should ignore and causing allergy," she says.

The hygiene **7**_____ has indeed changed, says Anna Fishbein, MD, an allergy and immunology fellow at Northwestern University and now an assistant professor at the University of Maryland. "It's become more complicated. It's no longer just getting exposed to the right number of microbes, but to good bacteria and viruses that alter the microbes in our intestines and protect us against both allergies and infections," says Fishbein.

But one child's good microbes are another child's bad **8**_____, DeMuth warns. "There is also an interaction between these microbes and an individual child's genetics," she says. "Certain people who have a dog in the house are protected against infections and allergies, but some are not. This is not a one-size-fits-all kind of thing." Worst of all, DeMuth says, is for the family of a sickly or asthmatic child to bring a dog into a pet-free home in the hope that it might help. "The absolute wrong thing is to put a dog in the house for kids with **9**_____," she says. "Yes, having a dog in the house early can protect against wheezing or respiratory infections. But this exposure has to happen very early in **10**_____."

A Better Flu Shot May Be on the Horizon (3)

Directions: Fill in a word that best completes each sentence.

Purvesh Khatri, of Stanford University, and his colleagues, had been examining the immune system for clues about how best to ward off the flu virus. They scanned the entire human genome for the genes that might be protective, but they weren't able to find any that would focus specifically on influenza. They next looked at the immune cells whose job is to fight incoming microbes, such as the **1**_____.

The researchers recruited 52 people for their study. They analyzed their **2**_____ systems and had them inhale flu viruses to see if they became infected. The researchers learned that the people who experienced the worst flu symptoms had lower levels of certain kinds of immune cells known as natural killer (NK) cells, while those who didn't get as sick seemed to have higher **3**_____ of NK cells. Khatri and his team traced the NK **4**_____ back to the gene that codes them, and the protein called CD94 that the gene produces. The more CD94 people had in their blood, the milder their flu, while people with lower levels of **5**_____ seemed to experience worse **6**_____.

Because higher levels of NK cells seem to indicate a heightened level of immune alertness, those cells are more likely to mount an effective defense against flu. That could provide the foundation for a more effective universal flu vaccine. Such a **7**_____ might trigger a stronger immune response than current vaccines. Current vaccines are redesigned yearly to match the strains of the flu that are circulating at that time.

"Until now we have been focused on creating vaccines that give [specific] immune responses. But this finding means that even before that virus is recognized by any immune cells, the baseline immune status of the natural killer cells seems to be a major player in deciding if somebody becomes infected or not. So now the question becomes whether we can design a vaccine that can move people toward this higher baseline status where they have a slightly higher proportion of natural **8**_____ cells," says Khatri.

According to these early study results, people who seemed to experience fewer flu symptoms had about 10% to 13% of their immune cells made up of **9**_____ cells, while those who had the worst symptoms had less than 10% of NK cells. While having a higher proportion of primed killer cells may be good for fighting the flu, it is still not known whether it could have negative health effects, as well. Overactive immune responses can be deadly, resulting in inflammation, immune cells destroying healthy tissues, and autoimmune diseases. "The big unknown is what the downside could be," says Khatri.

The results, while preliminary, and only in a limited number of people, are an important first step in identifying NK cells as a possible new target in fighting flu. Manipulating the immune **10**_____ has previously provided new ways of treating cancer, those lessons might be applied to make flu treatments more effective, as well.

Body Hair Is Natural (4)

Directions: Fill in a word that best completes each sentence.

Today, the regular removal of body hair is ubiquitous: More than 99 percent of American women rid themselves of body **1**_____. Men too have been getting in on the act, hence the rising popularity of the "back, sack and crack" waxing technique. The hairy chests and Playboy bushes of the 1970s are gone.

Yet the image of the "ideal body" as hairless is relatively recent. It wasn't until the late 1800s that non-native American women became concerned with body hair. Rebecca Herzig, author of Plucked: A History of Hair Removal, says that when Darwin's theory was popularized by the press, the characteristic of **2**_____ hair was transformed into a question of competitive selection. Herzig writes that hair was associated "with 'primitive' ancestry and an atavistic [ancient] return to earlier, 'less developed' forms."

Herzig makes the distinction that, in the 1800's **3**_____ were supposed to be hairy, and women were not. Hairiness in women became indicative of deviance. Researchers of the time associated excessive facial hair in women with insanity, she writes in her book. By the early 1900s, unwanted hair was a source of discomfort for **4**_____ who desired femininity. When hemlines rose, women took extreme measures to remove hair.

In the 1920s and '30s, women used pumice stones or sandpaper to depilate (remove hair), which caused irritation and scabbing of the skin. Some women tried modified shoemaker's waxes. Thousands were killed or permanently disabled by a cream made from the rat poison thallium acetate. It was successful in eliminating hair, and in causing muscular atrophy, blindness, limb damage, and death. X-ray hair removal was another treatment option in the late 1800s-early 1900s. Women would be **5**_____-rayed for three-four minutes as hairs withered away. The radiation caused scarring, ulceration, and cancer.

During World War II, there was a shortage of the thick stockings that women wore to cover their hairy **6**_____, and shaving—something that had previously been associated with men's routines—became a common practice for women. In the 1960s and '70s, doctors began prescribing hormonal **7**_____, like Aldactone and Androcur (which are now often used in male-to-female transitions), to combat hirsutism (male pattern hair growth in women). The side effects of this hormone therapy can include cancer, stroke, and heart attack, and its effectiveness in reducing hair growth is inconsistent.

Today, women still try to rid their bodies of hair with lasers, waxing, and bleaching, and depilatories (chemical hair removers). **8**_____ hair removal can cause severe burns, blistering, and scarring; **9**_____ is painful and unsanitary; **10**_____ can irritate and discolor skin; and depilatories can burn and scar. Banishing hair from the body means constantly needing shaving, plucking, waxing and lasering. Hair removal follows the common proverb: Beauty is pain. (Meanwhile, hair on one's head must be abundant, dyed, styled, weaved, extended and implanted.)

Celebrate Valentine's Day by Eating an Actual Heart (5)

Directions: Fill in a word that best completes each sentence.

Eating a heart might sound gross, especially if you're picturing a bloody sacrifice, but they can be quite delicious. **1**_____ are made of muscle, just like other meats, though the heart **2**_____ is structured differently so the texture is unlike a standard steak or chicken breast. Hearts are slightly tougher than normal meat because the heart gets so much use. The toughness is like a piece of chuck or a beef shoulder—you need to cook a heart right to keep it from being too **3**_____.

A chicken heart should be marinated before grilling or sautéing. A cow's heart is larger than a **4**_____ heart. You could cut a beef heart into strips to pan fry it, throw it on the grill, or braise it. Cutting it into strips will make it look less like an actual heart.

Organ meats in general have lots of vitamin B12, which you need to keep your blood cells and neurological system healthy. Hearts also have plenty of iron and zinc, without the added risk of giving you too much vitamin A. Beef livers contain a lot of vitamin A. (An overdose of vitamin **5**_____ can impair your own liver function.) Another benefit of eating organ meat is there's also a lot of protein, though they also tend to be high in cholesterol.

Organ meats are inexpensive! Most people pay top dollar for a nice steak because they don't want **6**_____ meat. Back during World War II—when the U.S. was worried about protein rationing—the Committee on Food Habits came up with a scheme to get people to eat more organ **7**_____. They figured out that women, who made most of their family's food decisions, weren't buying livers and hearts because they thought these cuts were only suitable for the lower classes—or for feeding to pets and livestock. Lecturing housewives on the benefits of organ meat wasn't enough—when the government tried, only about three percent of **8**_____ felt convinced to serve organs at home. Instead, the Committee on **9**_____ Habits got the ladies to talk through their issues with organ meat amongst themselves. Once the housewives felt they were a part of the decision, about a third of the wives wanted to serve their **10**_____ wholesome hearts.

Take a Deep Breath...Hold it... and Let Go...You Should Feel Better Now- - Science Says So! (6)

Directions: Fill in a word that best completes each sentence.

Joining the breath to the body is taught in yoga, meditation is about calming the breath, and people with panic attacks are told to reduce their breathing rate to reverse their anxiety symptoms. The part of the nervous system that controls our **1**_____ rate also sets the tempo of our arousal and mood.

The drive to breathe originates in a structure called the pre-Botzinger complex. This is a group of a few thousand nerve cells in the brain's medulla, which acts as a go-between linking the spinal cord with the forebrain. These nerve **2**_____ fire off a rhythm of pulses that drive respiration. Internal and external stimuli, like the detection of low blood oxygen or exercise, act on this rhythm of pulses and make you increase your breathing **3**_____ when you need to.

Stanford scientist Kevin Yackle and his colleagues have shown that about 200 of the cells in the **4**_____ complex seem to play a key role in linking your mood to your breathing **5**_____. Their study, published in the journal, *Science,* involved using a toxin to selectively remove a small amount of the pre-Botzinger **6**_____ cells from a group of mice. The researchers expected this to result in the **7**_____ being incapable of breathing. But, to their surprise, the animals appeared to be healthy and normal.

A closer inspection, however, showed that their breathing pattern had changed significantly. The toxin-exposed mice took more slow breaths. The also spent less time actively exploring their environments, devoted triple-time to grooming and double-time to sitting calmly. Measurements of their brain activity also showed a corresponding increase in theta waves, which are associated with a state of relaxation. When they needed to, however, the animals could still breathe rapidly, it was just that more stimulation would be needed to stir them from their calm state.

To understand why the removal of a small number of **8**_____ cells should make such a large difference, the researchers followed the nerve connections normally made by the group of cells they had removed. These cells, they found, use a nearby relay station in the brainstem, called the locus coeruleus, to send alerting signals to other parts of the **9**_____. These cells, the researchers speculate, link mood and arousal to breathing, perhaps to help us to prepare for extreme demands with our reflexes. They pick up on internal and external stimuli like low blood oxygen or exercise and, when breathing quickens in response to demands placed on the body, the nerves communicate with other body organs like the heart and muscles to deal with the demands being placed upon the body.

Yoga instructors are right to emphasize the importance of breathing. This is true also for a person hyperventilating during a **10**_____ attack: the best thing to do is to realize that rapid breathing is probably making the problem happen in the first place...

Dehydration in the Human Body (7)

Directions: Fill in a word that best completes each sentence.

Water makes up about 55 to 65 percent of your body. It's a crucial ingredient in the chemistry that helps your brain think, your blood flow, and your muscles move. Without water, the human body becomes dehydrated. The rate of dehydration is different for everybody because it is dependent upon the individual's amount of exercise, the temperature of the individual's environment, and the amount an individual typically sweats. Still, **1**_____ can get dangerous quickly.

The first stage of dehydration is thirst. The sensation of thirst is a register of a water loss of two percent of body weight. For a 170-pound person, that's 3 pounds. A person might lose this much sweat by kickboxing for an hour in a hot room without a drink.

When thirst kicks in, the human body clings to all remaining moisture. The kidneys send less **2**_____ to the bladder, darkening the urine. Sweating less, results in a body temperature increase. The blood becomes thicker and slower. The heart rate increases to maintain oxygen levels.

If water is not replenished and **3**_____ loss is increased to four percent of **4**_____ weight, fainting will ensue in the second stage. For a 170-pound person, that represents a 7 pound water loss. This is equivalent to riding a bike for three hours in extreme heat without rehydrating or going without water for two days.

At this point, the blood is so concentrated that the resulting decrease in blood flow makes the skin shrivel. Fainting occurs due to lowered blood pressure. The body has stopped sweating, and, without this coolant, the body overheats.

If water **5**_____ continues to seven percent of body **6**_____, organ damage can result. In a 170-pound person, that would be 12 **7**_____. You might lose this much sweat doing hot yoga for eight hours without rehydrating.

In this third **8**_____ of dehydration, the body is having trouble maintaining blood pressure. To survive, it slows blood flow to nonvital organs, such as the kidneys and gut, causing damage. Without the kidneys filtering blood, cellular waste builds up. The body is literally dying for a glass of **9**_____.

The human body can't continue to live once more than ten percent of it body weight in water is lost. If a 170-pound person lost 17 pounds of water, they would experience the final stage of dehydration: death. This is like going for five days, or running for 11 hours in 90-degree weather, without **10**_____.

In hot weather, uncontrollable body temperature means vital organs risk overheating. Liver failure will probably kill most humans. If temperatures are mild, toxic sludge would build up in the blood, and death would result from kidney failure. (Thirsty?)

Are e-Cigarettes and Vaping Safer Than Traditional Cigarettes? (8)

Directions: Fill in a word that best completes each sentence.

A decade ago e-cigarettes were unknown, but now they're mainstream, and usage is rising fast. Two independent studies have attempted to probe the question of whether e-cigarettes are safer than traditional combustible cigarettes. Based at UCL, Lion Shahab compared the chemical profiles of saliva, urine, and breath samples from just under 200 current smokers and "non-smokers", who were, by then, either exclusive e-cigarette or nicotine replacement therapy (NRT) users.

The average age of the participants was 37 years and the "non-smokers" were required to have quit more than 6 months before the study began. Across the three groups, nicotine levels were similar, showing equivalent levels of **1**_____ intake. But compared with current smokers, who had not **2**_____ for an hour before the tests were conducted, samples from the e-cigarette and NRT users contained 97% lower levels of chemicals like N-nitrosamines and carbonyls, which are known to cause cancer.

Commenting on the results, published in *Annals of Internal Medicine*, Shahab affirms "this would greatly reduce their risk."

Substituting **3**_____ for combustible tobacco products (cigarettes, pipes, or cigars) seems like a safer option, but not all physicians view e-cigarettes so positively. In a separate study published in the journal, *Tobacco Control*, the University of Michigan's Richard Miech believes that e-cigarettes might be a gateway product that could turn a new generation of teens into smokers, ironically at a time when **4**_____ rates have hit an international all low.

Miech has been surveying hundreds of school children and then following them up a year later to discover how many are "vaping", if this is linked to a future smoking habit, and whether **5**_____ changes the teen's ideas about the health risks of **6**_____.

"In 2011, 1% of school kids vaped," says Miech. "By 2015, it was 16% of 12 grade students. It's exploding."

A follow up **7**_____ one year later showed that only 7% of those who had <u>not</u> vaped now smoked combustible **8**_____ while 31% of students who <u>had</u> **9**_____ smoked **10**_____ cigarettes one year later. "Kids were 4 times more likely to have smoked a cigarette in the following year if they had been vaping," warns Miech.

The Effect of Age on Eyesight (9)

Directions: Fill in a word that best completes each sentence.

As humans grow older, their eyes may feel "tired" or "old". In fact, there are age-related processes that do affect eyesight. The good news is that identifying and treating these **1**_____ can often result in the preservation or restoration of excellent vision for an entire lifetime.

A good understanding of vision loss requires understanding the basic structure of the eye. For this reason, it is useful to compare the **2**_____ to a video camera: As light enters the eye--or the **3**_____--it travels through four main structures. These are the cornea, the lens, the retina, and the optic nerve. An entering light signal can be degraded or distorted by any of those structures, resulting in poor vision.

The entry point for light is the cornea, the tissue in the front of the eye. Next, the light signal encounters the lens, which focuses it on the third structure: the retina. The **4**_____ is analogous to the film in the back of the camera. The retina is where **5**_____ is converted into a neural signal that is interpreted by the brain. Finally, the optic nerve, which carries these signals to the brain, functions like a cable that connects the video camera to the television screen. Age-related vision loss results from a problem with one of these structures.

The corneal surface must be smooth for a clear image. A thin layer of tears coats its surface. These tears are vital to maintaining the cornea's smooth **6**_____. Any condition that disrupts this tear film can break down the cornea. With age, the cornea's surface can be damaged by inflammation of the eyelids, which may produce fewer tears. Blurry vision can result from the drying; if chronic, the condition is called dry eyes. Inflammation can be decreased with warm compresses, artificial tears, antibiotics and immunosuppressants.

The lens is also subject to the aging process. At birth, our eye's lenses are transparent (clear) and pliable (bendable). With aging, two things happen to the lens: it becomes more translucent (clouds up) and it becomes less **7**_____. The clouding of the lens (cataract) results in diminished vision. Similarly, when the lens loses its pliability (presbyopia), the hardened lens becomes more fixed cannot focus as well. Treatments exist for both conditions: cataract surgery involves replacing the affected lens with an artificial one; reading glasses or bifocals can correct presbyopia.

Finally, the retina converts the light image entering the eye into a neural signal. This signal is then ultimately transmitted to the **8**_____ via the optic nerve. Although these structures don't deteriorate with **9**_____, they are the site of age-related diseases. Macular degeneration (loss of central retinal function) and glaucoma (damage to the optic nerve due to increased intraocular pressure) lead the list of diseases. Early detection of these **10**_____ can often prevent or minimize vision loss, particularly as new and improved therapies become available. That's why people need proactive ophthalmic care with an emphasis on the development of new therapies to continue to have great vision as they age.

Energy from Pee (10)

Directions: Fill in a word that best completes each sentence.

Ioannis Ieropoulos, a bioengineer in England, uses living organisms to solve engineering problems. He grows microbes on a mesh made from carbon fibers. There they form a biofilm. When they are exposed to urine, the bacteria digest the organic matter in this liquid waste, and release electrons. The negatively charged particles then travel through the carbon mesh to wires. Such electricity-producing set-ups are known as microbial fuel cells. They provide electricity and can clean up wastewater.

How much power a **1**_____ fuel cell can generate depends on its size (and the number of bacteria it houses). Ieropoulos and his team have been working to make the **2**_____ cells smaller and easier to carry. The engineers have shown that a stack of 40 microbial fuel **3**_____ can charge up a cell phone battery. They also have used the fuel cells to light up bathrooms. Additionally, the bacteria in these fuel cells are remarkably resilient, says Ieropoulos. Once they form a biofilm, they can last about 3 weeks without any additional **4**_____ added.

At Stanford, environmental engineer Craig Criddle is also working with electron-producing bacteria. His team has created a microbial battery. Like the microbial fuel cell, bacteria inside his team's new **5**_____ grow on carbon filaments. They attach themselves to the **6**_____ using nanowires. These tiny strings hold the bacteria in place. They also act as electrical wires to carry electrons away from the microbes.

Engineers place the bacteria-coated filaments inside a small container filled with wastewater. As the bacteria digest the waste, electrons pass through their nanowires into the anode (negative end of the battery). From there, they travel to an outside circuit where they can power something or charge a portable battery.

The microbial battery captures **7**_____ from waste more efficiently than a microbial fuel cell can. In the fuel cell, energy is lost when some electrons react with oxygen and hydrogen to make water, Criddle explains. But in the **8**_____ battery, the electrons are transferred to an electrode that prevents such reactions from occurring. When the electrode fills up with electrons, the battery stops working. To fix this, the researchers simply remove the **9**_____, releasing the excess electrons into the air. When reinserted in the microbial battery, electrons flow freely and the battery goes back to work.

The microbial battery probably won't be portable. It won't plug into a device the way a typical battery does, but it could charge those types of batteries. It also could be used to capture energy from the bottom of the ocean, because electricity-producing bacteria naturally live there.

The idea of powering lighting or laptops with pee makes most people shudder. They see germy, smelly wastes, not the resources they hold. Fortunately, non-squeamish engineers are willing to harness **10**_____ as sustainable sources of power.

Is Expired Sunscreen Better Than No Sunscreen? (11)

Directions: Fill in a word that best completes each sentence.

Sunscreens typically provide protection with active ingredients that absorb or reflect ultraviolet (UV) radiation, such as zinc oxide or titanium dioxide, and so-called "broad spectrum" products that block out two types of potentially damaging UV **1**_____ — UVA and UVB rays. Most **2**_____ will remain effective up to three years after the container is opened, unless the brand's expiration date says otherwise. However, storage in hot places or exposure to moisture can break down a sunscreen's components and reduce its effectiveness even before it's "officially" expired says Dr. Lauren Ploch, a dermatologist.

"Any ingredient in a personal care product, even inactive ones, like emulsifiers and preservatives, can degrade over **3**_____," Ploch explained. "This degradation is often accelerated by suboptimal storage conditions, so storing sunscreen in a **4**_____ car may make it ineffective even before its **5**_____ date." Expired sunscreen may be less effective at blocking UV rays, raising the likelihood of sunburn and an increased risk of skin cancer. But heavy creams, which generally provide better coverage and sun protection than gels or sprays, can still provide a shield between skin and **6**_____, even if the sunblock is expired, Ploch said. "Expired sunscreen may be better than no sunscreen, especially if the active ingredient is a physical sunblock like zinc oxide or titanium **7**_____," she said.

However, sunscreens can differ widely in their composition of active **8**_____ and inactive **9**_____, and the storage history of individual containers of sunblock can vary even more. Therefore, it's impossible to say for sure how effective an expired tube of sunscreen could be, and users would be far safer seeking an unexpired option if one is available, she said. "I recommend borrowing sunscreen from someone else on the beach or going to a nearby store to buy something. Even small convenience stores and gas stations stock sunscreens now," Ploch said.

If you're caught without any sunscreen whatsoever, fabric can provide some protection from the sun, especially if the **10**_____ is specially woven or treated to screen out ultraviolet rays, says Ploch. In clothing, a UPF (ultraviolet protection factor) rating describes how effectively clothing will protect you, much as SPF (sun protection factor) numbers represent the effectiveness of sunscreen. But even UPF-treated fabric doesn't provide 100 percent protection, "so it's important to wear both sunscreen and sun-protective clothing," Ploch said.

Comparing Food Allergies: Animals and Humans May Have More in Common Than You Think (12)

Directions: Fill in a word that best completes each sentence.

Diarrhea after a glass of milk, an itchy palate (roof of the mouth) after eating apples, swelling in the face after consuming chicken eggs or a severe asthma attack due to peanut dust are all signs of an allergy or food intolerance. These symptoms are not limited to humans. Other mammals such as dogs, cats and horses may show similar symptoms after feeding. The number of pets affected by food allergies and intolerances is similar to that of humans. A research group of the European Academy of Allergy and Clinical Immunology (EAACI) focuses on exactly this issue. The group has recently published a paper that sums up food **1**_____ and intolerances in both **2**_____ and **3**_____.

All mammals (including humans) can develop allergies because their immune systems are capable of producing immunoglobulin E (IgE). Lead author Isabella Pali-Schöll explains that normally these special antibodies (IgE) help defend parasites or viruses. They are also responsible for type I fast-acting allergy symptoms such as hay fever, allergic asthma, and anaphylactic shock.

There are also very common non-immunologic forms of food intolerance. Erika Jensen-Jarolim, from the University of Veterinary Medicine Vienna, and another of the paper's authors, said that the causes and symptoms of **4**_____ intolerance are similar in both animals and humans. Food intolerance and immune response are typically caused by milk proteins, certain wheat proteins, soy, peanuts, tree nuts, fish, eggs and meat. The type of reactions found in dogs, cats or horses differ, however, from that of humans. The allergic reactions in animals mostly affect the skin, followed by the gastrointestinal tract. Asthma or severe shock reactions found in humans have rarely been observed in **5**_____.

Knowledge about the molecules that make up the allergens and a thorough comparison of adverse food reactions in humans and animals offers insight into the risk factors for the development of the condition and can lead to improved recommendations for the prevention and treatment of **6**_____ food reactions in animals and humans. There are currently no therapies for the treatment of food allergies in humans and animals so avoidance of the allergens is the only way to manage these conditions.

Immunotherapy treatments that can be administered under the tongue or on the skin are currently being researched, but it will take several more years for any products to be available for use. All that people or animals can do now to deal with the **7**_____ is an "elimination diet". This consists of removing all sources of protein from for the patient's diet and then slowly re-introducing one "normal" **8**_____ item at a time. During this time the patient is observed for any allergic **9**_____. This diagnostic procedure allows an allergen-free **10**_____ to be tailored to the respective food intolerance, while avoiding unnecessary restrictions.

When HIV Drugs Don't Cooperate (13)

Directions: Fill in a word that best completes each sentence.

Two drugs are considered synergistic in their effectiveness when using them together is more effective than using each by itself. Put another way, one drug that is synergistic with another **1**_____ doesn't just have a beneficial function, it makes the second drug perform its function better.

Combinations of drugs that treat HIV sometimes act synergistically, but other times they do not. Researchers at Thomas Jefferson University have discovered why certain **2**_____ only cooperate sometimes. The paper describing their research was published in the Journal of Biological Chemistry.

Second-line HIV drugs, used after first-line treatments have failed, target different steps in the process by which HIV enters human T cells (special white blood **3**_____ that help fight disease). Those two types of drugs are called *co-receptor antagonists* and *fusion inhibitors*. *Co-receptor antagonists* attach to receptors (special molecules on the cell membrane) on host cells. *Fusion inhibitors* attach to a viral protein called gp41. Sometimes these two **4**_____ are synergistic (perform better together), but sometimes they display no synergy at all (do not perform better **5**_____).

Michael Root and Koree Ahn of TJU put different doses of a *co-receptor antagonist* and a *fusion inhibitor* into cells and viruses with slightly different DNA. They found that many different factors are important for determining whether there's a synergistic interaction between these two drugs. The first factor affecting synergy was how many *co-receptors* were on the host cells. This number varies greatly between patients. Patients with a lot of *co-receptors* on their T-cells would get the beneficial results of synergy between the two drugs; patients with lower levels of *co-receptors* would not have the **6**_____ results of synergy between the two drugs. A second factor was the strength of the binding between the *fusion inhibitor* and gp41, which varies depending on virus' **7**_____. If the virus' DNA made the *fusion inhibitor* attach very tightly, then the two drugs will act synergistically, but if the virus' DNA made the fusion inhibitor attach weakly, the two drugs would not act synergistically.

These two results suggest that variations in viruses and in patients need to be considered when predicting how well **8**_____ combinations will work in helping to treat patients with **9**_____. As viruses like HIV evolve, the individual, as well as combinations of drugs used to treat them may not be as effective. This is bad news for HIV patients because adding **10**_____ drugs to a treatment regimen is a way to work against drug resistance. When a virus starts to become resistant to drugs, the benefits of synergy are lost.

Local Honey Might Help Your Allergies—But Only If You Believe (14)

Directions: Fill in a word that best completes each sentence.

Eating local honey to prevent the springtime sniffles seems like it should work: local bees collect pollen, pollen gets into the honey, you get exposed to the allergens, and your body learns they're safe. The belief that this works is so widespread that a group of scientists decided it was worth testing.

First, they started with an open trial, where participants knew whether they're getting the real treatment or a placebo. Volunteers with seasonal allergies were told either to eat a tablespoon of honey every day or to eat honey flavored corn syrup. Those eating the **1**_____ reported fewer symptoms.

Next, the researchers progressed to a double-blind trial, one where the participants didn't know whether they were getting the real treatment or a **2**_____. In this study, the participants were divided into three groups: one got local honey, one a national pasteurized honey, and one the flavored corn **3**_____. They ate a full tablespoon every day. Out of 36 initial volunteers, 13 dropped out. Those who survived eating a **4**_____ of honey every day for 30 weeks mailed in journals regularly tracking their allergy symptoms. After 30 weeks, those who ate the honey were doing no better with their allergies than those who ate the corn syrup.

It is possible that the participants weren't eating enough honey. The scientists note in their paper that oral consumption of allergens has been shown to be an effective way to train the immune system not to overreact. It follows that the allergens in honey should help train people's bodies. This might be difficult to test because few people will be able to tolerate eating multiple tablespoons of honey every **5**_____.

One more promising study suggests that it might also be about the type of honey. Local honey will have a variety of pollen sources, each of which may not be enough to have substantial microbial communities to train the eater's **6**_____ system. Finnish researchers decided to test the effect of birch pollen honey—regular honey, but with added bee-collected birch **7**_____. Birch pollen is one of the dominant season allergy sources in Finland, so the scientists gathered volunteers who were allergic to the tree and prescribed them either regular honey or birch pollen-enriched honey. A third control group ate no honey. Those who got the extra **8**_____ pollen had significantly reduced symptoms and more symptom-free days, even more than those who got regular honey. The control group in this study did not get a placebo. They were simply advised not to eat any honey-containing foods during the study period. It's possible that both forms of honey produced a strong placebo effect. This is a limiting factor in the validity of this study.

Like all naturopathic remedies, taking honey for seasonal allergy **9**_____ may make a sufferer feel better, a placebo effect—but the placebo **10**_____ can be helpful. If you believe the honey helps, then the honey helps. All that matters in the end is that you feel better, and if eating a tablespoon of honey is what enables you to spend summer days outside in the grass, you should go for it.

How to Tell if You Really Have a Fever (15)

Directions: Fill in a word that best completes each sentence.

Everyone knows the number 98.6 degrees Fahrenheit is normal body temperature. Except it isn't. Human bodies are all different, which means that 98.6°F is not necessarily the perfect **1**_____ for any one person. The reason we use 98.6°F is because of the research done by German physician Carl Reinhold August Wunderlich. He was the first person to attempt to standardize body temperature. He reportedly recorded over one million temperature readings from 25,000 people. He concluded that the average **2**_____ temperature of a healthy adult was around 98.6 degrees. He also noted that that number can vary—from as much as 97.2° to 99.5°—depending on the time of day, whether a person was male or female, and how old they were. However, 98.6 **3**_____ stuck as the standard of a "normal" healthy **4**_____.

Even though **5**_____ degrees is the standard, that number has been questioned over the past few decades. Doctors generally agree that there is not just one "normal" body temperature for all adults. As Wunderlich and many modern researchers have noted, body temperature fluctuates. One difference among people is gender. Women tend to have slightly higher body temperatures than **6**_____ due to differences in hormones, fat storage, and metabolism. Another difference among people is age. Older adults tend to have a lower body temperature than their middle-aged and young counterparts. Yet another reason for fluctuation is time of day. Each person's body temperature changes throughout the **7**_____, reaching a maximum in the early evening and a minimum in the early morning. Body temperature can also increase during and after intensive exercise.

Given the amount of **8**_____ for each person, it's best to use 98.6 degrees as a flexible guideline. Measuring health using body temperature is best done by first taking a baseline measurement of your personal average body temperature. To find your personal **9**_____, measure your temperature daily over an extended period (the longer the better). For best accuracy, your temperature should be taken at the same time of the day with the same thermometer. That will give the most accurate reflection of your "personal normal".

People should not ignore changes in their own normal body temperature. A fever often means that a person's body is undergoing some type of immunological response, often from an infectious agent, like a cold or flu virus or a bacterial infection. If a fever is suspected, a comparison should be made of your unwell-feeling temperature with your baseline average. If that number is at least 2 degrees higher than your norm, then, by Centers for Disease Control standards, you have a fever.

Numbers, averages, and statistics are helpful, but they aren't everything. Knowing how your own body feels when it is well, compared to when it is sick is just as important. Aches and chills typically accompany a fever. If those symptoms are present, you most likely have a **10**_____, regardless of what the thermometer says. Know your average, but listen to your body, as well.

Human-Dog Bond Provides Clue to Treating Social Disorders (16)

Directions: Fill in a word that best completes each sentence.

The chemistry behind social behavior in animals could help scientists identify new ways of treating social disorders such as autism and schizophrenia. "We think that the genetic foundation for **1**_____ behavior is very similar in dogs and humans," said Professor Jensen from Linköping University in Sweden. He coordinates the EU-funded GENEWELL project. In their research on dogs, his team has focused on oxytocin, a chemical known as the love hormone, which strengthens bonds in animals and **2**_____.

Said Prof. Jensen, "People have looked at the effects of oxytocin levels on both physical contact and eye contact and there's no doubt that **3**_____ is a really important player in maintaining the cooperation and the contact between dog and owner." To investigate, the researchers sprayed dogs' noses with either saltwater or oxytocin, and then monitored how they responded to different tasks.

Many dogs that received oxytocin quickly turned to their owners for help when the tasks became too difficult. The **4**_____ itself wasn't the only factor at play. Within the brain, oxytocin must attach to a receptor to begin working. Dogs can possess one of several different variants of oxytocin **5**_____, depending on their genes. What the researchers noticed was that some dogs, with a particular kind of receptor, responded very strongly to the oxytocin spray. They turned to their owners for help a lot faster than dogs with other variants.

The researchers found five different genes that were most strongly associated with the dogs' help-seeking behavior. Prof. Jensen says that four out of these five **6**_____ are also associated with autism, ADHD and **7**_____ in humans. Because of this similarity between dogs and humans, dogs may be used as model animals for studying social impairments in humans.

The researchers involved in this study suspect that the oxytocin system needs another chemical to work properly. That **8**_____, dopamine, is known to be important in the reward system of the brain. The researchers think that both chemicals combine to encourage cooperative behavior.

Professor Diana Prata, from the Instituto de Medicina Molecular in Lisbon, Portugal believes that for pro-social behavior, you need both systems to interact. Prof. Prata's research is largely based on MRI (magnetic resonance imaging) scans. Her team views what happens in the brain when people play "the prisoner's dilemma" game, where people are asked to choose either to cooperate with or betray a partner. If they **9**_____, they each receive half of a reward. If one cooperates and the other does not, the one who does not will win all the reward, while the one who did cooperate gets nothing. The scientists are interested in which brain areas are illuminated when participants experience a reward or punishment. Oxytocin is then delivered to see if anything changes. This research will indicate whether drugs for social disorders should take a combined approach using oxytocin and **10**_____, which could lead to more effective treatments.

Being Hungry Shuts Off Perception of Chronic Pain (17)

Directions: Fill in a word that best completes each sentence.

Pain is important for survival. Without it, we could leave our hand on a hot stove, intensifying an injury. The burn would result in acute pain. **1**_____ pain is sudden and caused by a specific thing.

Another kind of pain is long-term, and is linked to things like headaches, arthritis, and back problems. Long-**2**_____ pain is called chronic and arises from inflammation. Inflammatory pain can be debilitating and costly, preventing us from completing important tasks. In natural settings, the lethargy (tired feeling) triggered by chronic pain could even hinder survival.

University of Pennsylvania neuroscientists have found that the brain has a way to suppress chronic pain when an animal is hungry, so that it can go look for **3**_____, but still allowing it to respond acute pain. "In neuroscience we're very good about studying one behavior at a time," says J. Nicholas Betley. "My lab studies hunger. In the real world, things aren't that simple. You're not in an isolated situation where you're only hungry." If you're an animal, it doesn't matter if you have an injury, you need to be able to go eat to survive.

Curious about how hunger may interact with the sensation of pain, the researchers observed how mice that hadn't eaten for 24 hours responded to either acute pain or longer-term inflammatory **4**_____. The researchers found that hungry **5**_____ still responded to sources of acute pain but were less responsive to **6**_____ pain than the well-fed control group.

The researchers identified the part of the brain that processes the intersection between hunger and pain: a group of neurons called agouti-related protein (AgRP) neurons. They found that stimulation of only about 300 AgRP **7**_____ acted to suppress inflammatory pain. Further experiments pinpointed the neurotransmitter, a molecule called NPY, that is responsible for selectively blocking inflammatory pain responses.

The researchers are excited by the potential clinical relevance of their findings. If their findings hold up in humans, this neural circuit offers a target for reducing the chronic/inflammatory pain that can linger after injuries. **8**_____ pain is currently treated with opioid medications, which also inhibit acute pain. Since acute pain is beneficial for **9**_____, the researchers want acute pain receptors to remain intact, while only targeting inflammatory pain.

The scientists will continue considering how different survival behaviors integrate in the brain and how the brain processes and prioritizes them. "We've initiated a new way of thinking about how behavior is prioritized," Betley says. "It's not that all the information is funneled up to your higher thinking centers in the **10**_____ but that there's a hierarchy, a competition that occurs between different drives, that occurs before something like pain is even perceived."

What's the Difference Between Indoor and Outdoor Allergies? (18)

Directions: Fill in a word that best completes each sentence.

Despite affecting 50 million Americans, allergies aren't well understood. Allergens that cause the immune system to over-react can range from sunlight to onions. The symptoms of an attack are just as varied. The allergy aisle is filled with different products that treat these immune system 1_____-reactions. The outsides of the product boxes claim, "All day relief!" "Non-drowsy!" "Indoor and outdoor!" This begs the question: Are indoor and 2_____ allergens really that different?

"Your immune system doesn't discriminate whether you are sensitive to a grass, tree or ragweed pollen, they may all trigger allergy misery," says Clifford Bassett, of Allergy & Asthma Care of New York. That is true for whatever allergens are inside your house, as well. Pets and dust mites can both produce allergy symptoms. Some 3_____ are just more prevalent outdoors, while others are found 4_____.

"It's fair to say that each allergen is uniquely different on its own," says Sarena Sawlani, medical director of Chicago Allergy & Asthma. "Mold, dust, and pollen for example can each trigger completely different allergic responses." Those responses all rely on the same basic mechanism inside our bodies: histamines. Each person reacts to allergens differently. Dog dander may cause tearing; pollens may cause sneezing. These aren't differences in how allergens act inside the body, but changes in how the 5_____ system reacts.

Drugs in the pharmacy aisle also don't treat allergens differently. Antihistamines are the pill-based medicines most Americans take. 6_____ prevent histamines from binding to receptor on cells in mucus membranes. If the histamines were to bind to the receptors on the cells of 7_____ membranes, that would trigger inflammation. Corticosteroids are the active ingredient in nasal sprays. They prevent histamines from releasing in the first place by blocking the influx of immune cells to mucus 8_____. Antihistamines and 9_____ are slightly different mechanisms, but they both block the same reaction. Since people react differently to different medications, finding the right one is a matter of trial and error. A combination of meds may even be necessary to fully treat allergies.

Paying attention to the onset of symptoms can help determine if allergies are coming from inside or outside the home. If symptoms are year-round, the allergen is probably inside the home; if symptoms peak during spring and summer, outdoor allergens are probably at least a partial factor. "Air pollutants can synergize with pollen in the air to create a double whammy effect and magnify and further worsen seasonal symptoms," explains Bassett. The only way to completely identify the allergens involved is to be tested by an allergist.

An allergist can determine if you are a good candidate for allergy shots, which can lessen allergic reactions by teaching the immune system not to treat allergens as dangerous invaders. If allergy shots aren't the best solution, an 10_____ can offer information about how to best handle symptoms.

Common Ingredient in Packaged Food May Trigger Inflammatory Disease (19)

Directions: Fill in a word that best completes each sentence.

Gut microbes help us fight off infections and resist allergies, but there's one thing we don't want them to do: touch our intestinal lining. Normally, a layer of mucus separates intestinal cells from gut bacteria. If **1**_____ reach these cells, they can stimulate the immune system and cause inflammation. This **2**_____ of the digestive tract is typical of inflammatory bowel disease (IBD), which causes diarrhea, fatigue, and abdominal pain. Chronic inflammation has also been associated with metabolic syndrome, a cluster of co-occurring conditions that increase a person's risk of heart disease and diabetes.

Benoit Chassaing, a microbiologist at Georgia State University, wondered if such a bacterial invasion could explain the correlation between the increasing use of food additives in industrialized countries and the incidence of IBD. Chassaing and his colleagues focused on emulsifiers—detergent-like compounds that make a smooth, creamy mixture of ingredients that would normally separate, like the milk fat and water in ice cream.

In this study, researchers fed two common emulsifiers to both a genetically susceptible mouse strain and wild-type mice—those without genetic mutations that would put them at increased risk of **3**_____ or metabolic syndrome. Among the susceptible mice, eating or drinking emulsifiers for 12 weeks increased the risk of developing symptoms of colitis (like intestinal inflammation seen in humans with IBD). The wild-**4**_____ mice showed low-grade inflammation in their intestines and some features of metabolic **5**_____: increased body fat, food intake, and blood sugar. These indicate poor glucose regulation associated with diabetes.

Microscopic imaging of the intestines revealed that the average distance between gut bacteria and the intestinal **6**_____ was reduced by more than half; bacteria seemed to be advancing toward the gut lining. Chassaing believes that emulsifiers damage the mucus layer directly, leaving it vulnerable to bacteria, or they change the composition of the microbiota, favoring the mucus-penetrating microbes.

It's not at all surprising that **7**_____ influence gut microbes, says Mia Phillipson, a physiologist at Uppsala University in Sweden who studies **8**_____ inflammation and bacteria-mucus interactions. "I think we're just on the verge of realizing how important this is," she says. As for the implications in humans, Phillipson says it's too early to make broad recommendations, but she suggests that people with **9**_____ or a family history of the disease consider avoiding these ingredients.

Chassaing, too, is careful not to cast emulsifiers as the ultimate villain. "Of course, society changed so drastically during the last century … and so many other factors were used in food that we can't really know yet which one is playing the most important role." His group is now preparing a more ambitious study that compares the microbiomes of people who completely avoid **10**_____ for several weeks with those on a standard Western diet. The current research finding has been published in *Nature*.

Students Know About Learning Strategies -- But Don't Use Them (20)

Directions: Fill in a word that best completes each sentence.

The first year in a university is a steep learning curve for many students. One big challenge is planning and organizing their own learning and dealing with various forms of academic assessment, from multiple-choice exams to essays.

New students typically work out their own strategies for **1**_____ through trial and error. However, strategies to prepare for one type of test or assignment may not work for **2**_____. As a result, students may find themselves underprepared. Even post-graduate students can encounter new challenges, such as writing a master's thesis, that might require different learning techniques.

Self-regulated learning (SRL) strategies are an effective way for students to maximize their academic potential and are considered essential for **3**_____ success by educational researchers. "SRL refers to evaluating, planning, and executing your own learning," says Nora Foerst of the University of Vienna. "SRL includes different learning **4**_____, such as planning your approach, structuring learning content, rewarding yourself after accomplishing a goal or making realistic demands to avoid frustration."

Previous studies found that many students know about common **5**_____ strategies. However, researchers are less sure how often students use the techniques, whether they can use them effectively, and whether they know which techniques are most appropriate in specific learning situations. Since these things are unknown, Foerst and her colleagues surveyed students enrolled in Psychology or Economics programs about their learning strategy knowledge and actions. The researchers asked **6**_____ whether they knew about helpful SRL strategies for specific learning situations. They also asked whether the students put the **7**_____ into practice, and if not, why not.

As expected, most students could correctly identify many SRL strategies. However, fewer students applied them while studying. In some cases, as many as one-third of the students who correctly identified a technique as beneficial admitted that they didn't **8**_____ it in their own learning. Both Psychology and Economics students showed a similar disconnect between knowledge and action. Psychology students were slightly better at identifying the strategies, likely because their curriculum included information about SRL techniques.

Foerst's survey revealed a variety of reasons for not using self-regulated learning strategies. Many students felt they didn't have enough time to use the strategies or were unable to apply them effectively. Some did not see the benefits of the strategies for specific tasks or believed that using them would be too much work. Foerst wants to encourage universities to provide more SRL training for their students and provide them with hands-on **9**_____ to learn how and when to apply SRL strategies for specific learning **10**_____. Students need help to understand that the techniques could save them time and enhance their learning outcomes.

Regular Physical Activity is 'Magic Bullet' For Pandemics of Obesity, Cardiovascular Disease (21)

Directions: Fill in a word that best completes each sentence.

The statistics on regular physical activity in the United States are bleak. Only about 20 percent of Americans engage in recommended levels of exercise. About 64 percent of Americans never do any physical **1**_____. The statistics are not much better in Europe. Only 33 percent of people engage in physical activity regularly. As many as 42 percent never do any **2**_____ activity.

"If regular physical activity were a pill, then perhaps more people would take it," said Charles Hennekens, senior academic advisor to the dean at Florida Atlantic University's College of Medicine. In an editorial published in *Cardiology,* professors from FAU have examined all the evidence and conclude that exercise is closest to a "magic bullet" to fight against the spread of obesity and cardiovascular disease worldwide.

Being overweight or obese in middle age increases the risk for cardiovascular disease, type 2 diabetes, osteoarthritis, and some types of cancer. Physical activity has benefits including lowering blood pressure and triglycerides, decreasing the risk of diabetes, heart attacks, and colon cancer. Physical activity may even decrease the risk of arthritis, as well as improve mood, energy, sleep and sex life.

With all these great benefits, why don't people exercise more regularly? The authors suggest that the time and effort needed for regular physical activity, along with the limited knowledge about the **3**_____ of exercise, may contribute to the sedentary (inactive) lifestyles of many people. This hypothesis is supported by the survey data from Europe where 42 percent give this as the main reason for their **4**_____ lifestyle.

According to Steven Lewis, co-author and professor at FAU, there are mistaken beliefs about the role of physical activity, caloric intake, and calories burned during exercise. As a result, calorie restriction dieting is seen as more practical for weight control than regular physical activity. Hennekens adds that most people have difficulty achieving and maintaining **5**_____ loss solely by restricting their **6**_____ intake. Modern inactive lifestyles seem to be as important as diet in causing **7**_____.

The authors note that brisk (fast) walking for only 20 minutes a day burns about 700 **8**_____ a week and results in a 30 to 40 percent reduced risk of heart disease. They stress that regular physical activity also should include resistance exercise such as lifting weights, which adds to lean body mass and promotes calorie burn at rest, which helps to control body weight. Both walking and **9**_____ exercise can be safely performed by patients with heart failure and the elderly.

"Clinicians and their patients should remain cognizant of the crucial role of regular physical activity in improving the quality and quantity of life," said Hennekens. "There is a dire need to educate patients about the importance of regular physical activity for weight **10**_____."

Our Muscles Measure the Time of Day (22)

Directions: Fill in a word that best completes each sentence.

Biological clocks are ticking everywhere throughout our body. They trigger the release of the hormone melatonin during sleep, favor the secretion of digestive enzymes at lunchtime or keep us awake at the busiest moments of the day. A "master clock" in the brain synchronizes all the body's glands and organs. Researchers in several institutions funded by the Swiss National Science Foundation (SNSF) have found that the same sort of biological or circadian clock is at work in our muscles. Their research shows that changes in this machinery might be important in the development of type 2 diabetes.

According to this research, different types of lipids (fats) are favored by cells, over other types, depending on the time of **1**_____. The international team of researchers tested the hypothesis that a biological clock regulated the lipid cycle in the muscles of volunteer research subjects. The **2**_____ had to follow a predictable schedule of eating and sleeping for one week to synchronize their master clocks. Then, every four hours, researchers would take a very small sample of thigh muscle tissue and analyze its lipid composition.

The research team observed a relationship between the variety of **3**_____ that make up muscle **4**_____ and the time of day, but they needed further evidence because the combination of lipids was very different from one research subject to another. To corroborate (further support) their findings, the researchers grew human **5**_____ cells in the lab and artificially synchronized them (put them on predictable schedules) using a signal molecule normally secreted in the body. The researchers then observed regular changes in the cell's lipid composition. When the researchers stopped using the predictable "clock" schedule, the cells stopped exhibiting regular changes in the composition of their lipids. These lab results matched the results found in the human **6**_____.

Both findings lead researchers to the conclusion that the variation of lipid types in the muscles of the research subjects is due to **7**_____ rhythm. First author Ursula Loizides-Mangold explains that "The main question is still to be answered: what is this mechanism for?" One possibility is that the biological clock in muscles could help in regulating the cells' sensitivity to **8**_____. Since lipids are components of cell membranes, they may influence the insulin molecules' ability to travel into and out of muscle cells. Changes in the cell membrane's lipid **9**_____ could make muscles sensitive to the insulin hormone, and, as a result, effect the cell's ability to take in blood sugar.

A low sensitivity of muscle cells to insulin leads to a condition called insulin resistance, which is known to be a cause of type 2 **10**_____. There is a strong link between circadian clocks, insulin resistance, and diabetes development. New lab technologies involving how muscle cells are studied will aid in the further investigation of the hypothesis that lipid metabolism is linked to circadian rhythms and type 2 diabetes. When this hypothesis is further supported new treatments could be developed.

Tissue-Engineered Human Pancreatic Cells Successfully Treat Diabetic Mice (23)

Directions: Fill in a word that best completes each sentence.

A study published by *Cell Reports* describes a bioengineering process called a self-condensation cell culture. The technology helps nudge medical science closer to growing human organ tissues from a person's own cells for regenerative therapy, say study investigators.

"This method may serve as a principal curative strategy for treating type 1 diabetes, of which there are 79,000 new diagnoses per year," said Takanori Takebe, of Cincinnati Children's Center for Stem Cell and Organoid Medicine. "This is a life-threatening disease. Developing effective and possibly permanent therapeutic approaches would help millions of children and adults around the world."
1_____ is the study's co-lead investigator along with YCU colleague, Hideki Taniguchi.

Scientists tested their processing system with donated human organ cells (pancreas, heart, brain, etc.), with mouse organ cells, and with induced pluripotent stem cells (iPS). Reprogrammed from a person's adult cells (like skin cells), iPS cells act like embryonic cells and can form any tissue type in the body. The tissue-engineering process also uses two types of embryonic-stage progenitor cells, which support formation of the body and its specific organs. The progenitor **2**_____ are mesenchymal stem cells (MSNs) and human umbilical vascular endothelial **3**_____ (HUVECs).

Using either donated organ cells, mouse cells or **4**_____ cells, the researchers combined these with MSNs, **5**_____ along with other genetic and biochemical material that cue the formation of pancreatic islets. In conditions that nourish and nurture the cells, the ingredients condensed and self-organized into pancreatic islets. Those **6**_____ islet cells were then transplanted into humanized mouse models of severe type 1 diabetes. There, they resolved the animals' disease, report researchers.

Human pancreatic **7**_____ can already be transplanted into diabetic patients for treatment. Unfortunately, the success rate is low because the tissues lose their vascularization (blood supply) as the islets are being processed before transplant. "We need a strategy that ensures successful engraftment through the timely development of vascular networks," said Taniguchi. "We demonstrate in this study that the self-condensation cell culturing system promotes tissue vascularization."

The genetically engineered pancreatic islets quickly developed a vascular network after **8**_____ into animal models of type 1 **9**_____, as well as functioned efficiently as part of the endocrine system -- secreting hormones like insulin and stabilizing glycemic control in the animals. Previously, Takebe's and Taniguchi's research team had demonstrated the ability to use iPS cells to tissue engineer human liver organoids that can vascularize after transplant into laboratory mice. Now, these scientists have shown the ability to generate organ tissue fragments that **10**_____ in the body in pancreatic islets.

Six Genes Driving Peanut Allergy Reactions Identified (24)

Directions: Fill in a word that best completes each sentence.

Mount Sinai researchers have identified six genes that activate hundreds of other genes in children experiencing severe allergic reactions to peanuts. This is the first study in human trials to identify genes driving acute peanut allergic reactions using a double-blind placebo-controlled approach with comprehensive sequencing of genes expressed before, during, and after they ingested peanut. The study was also the first to study gene expression in children over the course of their allergic **1**_____, allowing each subject's reaction to be compared to their own pre-reaction state, rather than to a control group without peanut allergies. This approach allowed researchers to accurately detect **2**_____ expression changes resulting from the reactions.

The results of the study have been published *Nature Communications*. The standard treatment for individuals with peanut allergy includes avoidance of peanuts and prompt care for the **3**_____ reaction. Immunotherapy has demonstrated progress, but it is not effective for all individuals, carries adverse side effects, and has not been approved by the U.S. Food and Drug Administration.

"This study highlights genes and molecular processes that could be targets for new therapies to treat peanut-allergy reactions and could be important to understanding how **4**_____ allergy works overall," said the study's senior author, Supinda Bunyavanich, MD, MPH, Associate Professor, Pediatrics and Genetics and Genomic Sciences at the Icahn School of Medicine at Mount Sinai.

The research team collected blood samples from 40 peanut-allergic children before, during, and after a randomized, double-blind, placebo-controlled oral food challenge. Subjects ingested incremental amounts of peanut at 20-minute intervals until an allergic reaction occurred or a cumulative dose of 1.044 grams of peanut was ingested. In a similar fashion on a different day, the same subjects **5**_____ incremental doses of placebo oat powder; again, blood samples were drawn before, during, and **6**_____ the challenge. The team then performed comprehensive RNA sequencing on the **7**_____ samples followed by computational data analyses to determine what genes and cells were activated and driving these allergic reactions.

"Other studies have looked at genes expressed in people with food allergies and compared them to people who don't have **8**_____ allergies," said Dr. Bunyavanich. "One of the strengths of our study is that we looked at genes expressed over time in children actively reacting to peanut and followed that person throughout their reaction, which provided a detailed and comprehensive picture of what's happening on the genetic and molecular level during a **9**_____-allergy reaction."

One of the limitations of the **10**_____ is that it only focused on peanut allergy. Dr. Bunyavanich and the research team plan to conduct future studies targeting other common allergens, such as milk and egg, to address if their findings may be relevant to other types of food allergy.

What the Consistency of Your Poo Says About Your Health (25)

Directions: Fill in a word that best completes each sentence.

Our gut does more than help us digest food; the bacteria that live in our intestines have been implicated in everything from human mental health to weight gain. In fact, the feces (or stools) humans produce can provide a window into the health of the large intestine (or colon) and overall gastrointestinal tract.

In 1998, Stephen Lewis and Ken Heaten from the University of Bristol developed a seven-point stool scale, ranging from constipation (type 1) to diarrhea (type 7). Today, the Bristol Stool Chart allows people with gastrointestinal symptoms to clearly describe what they are seeing in the toilet without having to provide samples. For most people, the form of stool that is excreted can vary widely depending on what has been eaten, how much the body has exercised, or when the last bowel movement occurred. Ideally, stools should be easy to pass without straining and without any sense of urgency. Poo on the Bristol **1**_____ Chart describes normal poo as smooth or cracked, and sausage or snake shaped.

The drier stool tends to compact and can apply long term pressure to the large **2**_____. This can result in lacerations, hemorrhoids, infection, or inflammation. Watery stool forms may be associated with gut infections. As a rule, softer but not watery stool forms are best. A change of bowel habit that leads to the sustained production of constipation or diarrhea should be discussed with a doctor.

Drinking enough water is important for bowel health. Water recycling is one of the key functions of the **3**_____ intestine. The human body invests 9 liters of fluids from saliva, stomach secretions, and bile into digestion daily. Yet the body doesn't defecate that volume of fluid. The longer it takes for digested **4**_____ to pass through the large intestine, the more water gets reclaimed and the drier the stool becomes. Factors affecting the transit rate of food through the gastrointestinal **5**_____ have significant influence on stool form.

Antibiotics, pain killers, and physical inactivity all reduce gut contraction, which slows the passage of food through the large intestine. This can lead to constipation. Diets also play a role in stool form and health. **6**_____ is more common in people consuming Western-style diets. This is associated with increased incidence of colon cancer, inflammatory bowel diseases, as well as other chronic diseases. **7**_____-style diets tend to have less fiber. Fiber impacts transit time and stool **8**_____.

A healthy, well-hydrated person eats dietary fiber and roughage. The food takes up water and swells. This increases the volume of the stool, softening it, stimulating more rapid transit. Americans who eat a Western-**9**_____ diet may consume as little as 12-15g of fiber per day. Consumption of at least 30g of dietary fiber per day is recommended. However, if you have gastrointestinal symptoms, fiber may not help. You may need to carefully consider the type of fiber you consume. For most of us, though, more **10**_____ in our diets reduces food transit times, softens stools, makes bowel movements more comfortable and improves intestinal health.

A New Reason Why Newborns Can't Fight Colds (26)

Directions: Fill in a word that best completes each sentence.

Newborns are more likely than older babies to catch, and die from, serious infections. The reason is unclear. Indeed, there may be more than one explanation. One theory is infants' immune systems haven't fully matured. Another is that both mothers-to-be and their fetuses in-utero have suppressed immune **1**_____, so that neither rejects the other. After birth, it takes babies a month or so to boost their immunity.

Seeking new ways to better understand this process, Sing Sing Way, a pediatric infectious disease doctor, wondered whether transferring immune cells from adults might speed the development of newborns' **2**_____ systems. Yet when he and his colleagues injected infection-fighting cells from the spleens of adult mice into 6-day-old pups, nothing happened: The pups were just as vulnerable to harmful bacteria as control animals. Way's group later did the reverse transplant where they gave adult **3**_____ newborn immune cells that were inactivated in the pups and found that they "turned on" in the mature animals. These experiments "told us it wasn't a problem with the neonatal cells themselves," Way says. Rather, he believes, the environment (either the newborn's or adult's body) guided how the **4**_____ behaved.

Researchers in Way's lab next looked to the gut microbiome, the constellation of healthy bacteria that populates human intestines as a possible reason as to why newborns get more serious **5**_____. Newborn mice, just like human babies, are born "clean," with little intestinal bacteria. That changes rapidly. Way wondered whether there might be a connection between colonization and a purposeful suppression of the immune system in his mice. Way's team discovered the **6**_____ had many immune cells that expressed a surface receptor called CD71, which causes immune suppression of other cells. As the mice grew older they had fewer and fewer cells with CD71 receptors. Way believes that, as the intestines of the mice were colonized by bacteria, less immune **7**_____ was necessary by their bodies.

The research, reported in *Nature*, "adds a new and very important chapter" to the story of how the immune system develops, says Mike McCune, an immunologist at the University of California. Immune suppression, at least in newborn mice, appears to reflect a purposeful shift in the balance of immune cells. One big caveat is whether what Way's team observed will hold up in human **8**_____. A baby's immune system develops differently than that of a mouse.

Way is interested in looking for CD71 cells in babies born around their due date, as well as those born prematurely. Very premature infants often die of a condition called necrotizing enterocolitis, a massive inflammation of the intestines. That may be driven partly by a lack of **9**_____ cells in these babies, if their immune systems are still fetal and not ready for the natural colonization of gut bacteria that happens after they're born. If so, perhaps, in the future, babies born **10**_____ could receive immune cells that would make their immune system more like a full-term baby's, allowing their guts to stay healthy.

Here's What Happens to Your Body If You Eat Too Much Salt (27)

Directions: Fill in a word that best completes each sentence.

Salt does more than just make your food taste more delicious—it's important for your body to function properly. Sodium, one of the key ingredients in table **1**_____, regulates blood flow and pressure, and helps transmit messages between nerves and muscle fibers. Chloride, the other chemical in **2**_____ salt, aids in digestion. Foods in your diet need to have enough salt to replenish these nutrients to keep you healthy.

Too much salt can be bad for you, however. Processed foods are packed with salt; restaurants add salt to their food to make it **3**_____ better. As a result, more Americans are eating high-sodium diets, which has drastic effects on their health.

More salt in the **4**_____ means the kidneys keep more water in the body. Kidneys, which filter out waste from the blood, maintain a special balance of sodium, potassium, and water. The consumption of excess **5**_____ makes the body hold extra water which has lots of undesirable effects, such as swelling in the hands, arms, feet, ankles, and legs. More fluid in the body means more **6**_____ coursing through veins and arteries, which, over time, causes them to stiffen, which could lead to high blood pressure.

Salt makes you thirsty, which is the body's way of trying to correct that sodium-water balance. Drinking lots of **7**_____ can worsen swelling and blood pressure. However, not drinking enough could force the body to draw water out of other cells, causing dehydration.

People who consume high-sodium diets usually urinate more because of all the excess water. Calcium is lost with each urination. Calcium makes strong bones and teeth and if excess **8**_____ causes calcium loss then bones may be weakened, causing osteoporosis.

There may also be additional negative effects still not fully understood: excess salt may cause stomach ulcers, infections, and may even hasten stomach cancer because sodium may disrupt the stomach's mucus lining.

The evidence is clear: Too much salt can have serious long-term health implications. Still, lots of people eat **9**_____ where the sodium intake exceeds the daily-recommended value of 2,300 milligrams.

To better inform the public, some **10**_____ mark dishes on their menus that exceed the daily-recommended sodium intake. Hopefully, if people are more aware of the salt they consume, they will adjust their diets accordingly, and be healthier as a result.

Self-Esteem in Kids: Lavish Praise Is Not the Answer, Warmth Is (28)

Directions: Fill in a word that best completes each sentence.

Who am I and what is my place in the world? Children are born without an answer to these pressing **1**_____. As they grow, they recognize themselves in the mirror, refer to themselves by their own name, evaluate themselves through the eyes of others and understand their standing in a social group.

Research by Christina Starmans from the University of Toronto shows that even toddlers have an idea of what it means to have a 'self'. Young children see the **2**_____ as something that is unique to a person, separate from the body, stable over time, and located within the head, behind the eyes. Research by Andrei Cimpian (New York University) and his colleagues shows that even toddlers have the cognitive (mental) ability to form self-worth (to know how satisfied they are with themselves as people).

Over time, noticeable individual differences arise in children's self-concept. Some children like themselves, whereas others feel negatively about **3**_____. Some children see themselves as superior and deserving special treatment, whereas others consider themselves to be on an equal plane with others. Some children believe they can grow and build their abilities; others believe their abilities are fixed and unchangeable. Where do these individual differences come from? What leads children to see themselves the way they do?

4_____ form their self-concept, at least in part, based on their social relationships. For example, research by Michelle Harris (University of California) and her team shows that children develop higher self-esteem when they receive warmth from their parents. Warm **5**_____ show an interest in their children's activities and share joy with them, which makes children feel noticed and valued. Children may develop lower **6**_____ and sometimes even narcissism (extreme selfishness) when their parents give them lots of extremely positive, inflated praise, such as "Wow, you did incredibly well!" Such inflated praise may give children a sense of grandiosity (an exaggerated belief in their own importance), but at the same time also make them worry about falling short of the standards set for them.

Previous research has shown the importance of having a growth mindset - the belief that you can develop your skills through effort and education. Children with a **7**_____ mindset are eager to take on challenges, persist when the going gets tough, and see failure as opportunities for growth. Kyla Haimovitz and Carol Dweck (Stanford University) describe how parents can foster a growth **8**_____ by praising children for effort instead of ability (for example, 'You worked so hard!') and by teaching **9**_____ that failure isn't harmful but benefits learning and growth. Parents can encourage children to ask themselves: why did I get such a low grade, and what can I do differently in future?

This research shows that children construct their **10**_____ based on the social relationships they have, the feedback they receive, the social comparisons they make, and the cultural values they support.

Smartphone-Controlled Smart Bandage for Better, Faster Healing (29)

Directions: Fill in a word that best completes each sentence.

Researchers at Harvard and Tufts Universities have developed a "smart bandage" capable of electronic delivery of wound treatment. Existing bandages range from basic dry patches to those with embedded medication which can only treat a wound with simple diffusion. The smart bandage consists of electrical fibers coated in a gel that can be loaded with infection-fighting antibiotics, tissue-regenerating growth factors, painkillers or other medications. A microcontroller no larger than a postage stamp, which could be triggered by a smartphone or other wireless **1**_____, sends small amounts of voltage through a chosen fiber. That **2**_____ heats the **3**_____ and its surrounding gel, releasing treatments it contains. A single bandage could accommodate multiple medications tailored to a specific type of wound, while offering the ability to control the dose and delivery schedule of those **4**_____. That combination of customization and control could accelerate the healing process. The new smart bandage was described in an article in the journal Advanced Functional Materials.

The researchers envision the smart **5**_____ being used to treat chronic skin wounds that stem from diabetes. More than 25 million Americans may suffer from such wounds in the future. The CDC estimates that diabetes cases could double or triple by the year 2050. There are large medical costs from these types of wounds, so this type of bandage could help manage that **6**_____.

Those wounded in combat might also benefit from the bandage's versatility and customizability. Soldiers on the battlefield may be suffering from several different injuries or infections. They might be dealing different types of pathogens, depending on their environment. The smart bandage is variable and can have antidotes or drugs targeted toward specific hazards in the **7**_____.

To evaluate the potential advantages of their smart bandage, researchers ran a series of experiments. In one, the scientists applied a smart bandage loaded with growth factor to wounded mice. When compared with a dry **8**_____, the team's version regrew three times as much of the blood-rich tissue critical to the healing process. Another experiment showed that an antibiotic-loaded version of the smart bandage could kill infection-causing bacteria. The **9**_____ also demonstrated that the heat needed to release medications did not diminish the medicine's strength.

Though the researchers have patented their design, it will need to undergo further animal and then human **10**_____ before going to market. That could take several years, though the fact that most of the smart bandage's components are already approved by the Food and Drug Administration should streamline the process. In the meantime, the scientists are also working to incorporate thread-based sensors that can measure glucose, pH and other health-related indicators of skin tissue. That capability would allow the team to create a bandage that could automatically deliver proper treatments.

Does Your Back Feel Stiff? Well, It May Not Actually Be Stiff, Study Finds (30)

Directions: Fill in a word that best completes each sentence.

"My back feels so stiff!" We often hear our friends say. That doesn't actually mean your friend's back is stiff like a board, according to a new study at the University of Alberta's Faculty of Rehabilitation Medicine.

"A conscious experience of feeling **1**_____ does not reflect true biomechanical back stiffness," explained Greg Kawchuk, professor and back and spine expert in the Department of Physical Therapy. "When we use the same word, **2**_____, to describe a feeling and how we measure actual stiffness, we assume these words are describing the same **3**_____. But that is not always the case."

In the study, Kawchuk and his team asked participants how stiff their **4**_____ felt to them. After that, using a customized device, they **5**_____ just how inflexible (or stiff) the back was, in reality. "There was no relation between biomechanical stiffness and the reported **6**_____ of stiffness," he said. "What people describe as stiffness is something different than the **7**_____ of stiffness."

Tasha Stanton, lead author and senior research fellow of pain neuroscience at the University of South Australia, said that the feeling of stiffness may be a protective thought or idea that is created by our nervous system. "It's our body's way of protecting ourselves, possibly from strain, further injury or more pain," she said.

With lower back pain being the leading cause of disability worldwide affecting approximately 632 million **8**_____, it is important to examine processes associated with lower back **9**_____ and its symptoms, including stiffness.

Words are important. The words patients use to describe a problem in the clinic may not be the same thing as medical staff measure in the **10**_____. "We need to find out what it means exactly when someone says they have a stiff back. We now know it might not mean that their back is mechanically stiff," said Kawchuk, "It could mean they feel their movements are slower and more painful."

Is It Possible to Eat So Much That Your Stomach Explodes? (31)

Directions: Fill in a word that best completes each sentence.

You might think your stomach is like a balloon: Both start out small when empty; Both get bigger when stuff goes in them. But this is where the comparison ends.

If you blow up a **1**_____, pressure and volume have a fixed relationship: as pressure goes up, so does **2**_____. In a balloon, **3**_____ and volume have a direct relationship. In the stomach, there is not the same **4**_____.

The state of the stomach is determined not only by what you put into it. Instead, it's controlled by nerve inputs and hormones. The effect of hormones on the gut is difficult to measure, but nerve **5**_____ are more straight-forward.

A resting, empty stomach can hold six and a half to just over 10 fluid ounces. That is less than a can of Coke. That stomach volume can more than double the instant you start eating. It can even double if you're just salivating and thinking about **6**_____. A brainstem reflex through the Vagus nerve tells the stomach that food will soon be coming so the stomach relaxes to make room.

Unlike a balloon, the increase in pressure from food or liquid doesn't cause the stomach to relax and expand. The stomach expansion is anticipatory. It is the result of your brain talking to the nerves in your stomach. Your **7**_____ is anticipating a meal.

This expansion isn't at all like a balloon filling with air; it's more like an unfolding. The stomach is covered in rugae (folds) that unfurl as the stomach organ grows. The rugae have limits. A layer of nerve cells lines the walls of your stomach. Once you've reached a point of too much **8**_____ or drink, those nerves send a signal to your brain telling it to stop eating.

Although the stomach size can grow, there is also an "exit hatch". It only takes about 10 to 15 minutes before all that food and drink that was eaten begins to pass through the "**9**_____ hatch". How big a stomach can grow in 10 to 15 minutes depends on factors such as age, health, eating habits, and hormones. Of course, this depends on a person listening to what their stomach tells them.

For most people, that "Ooof, I ate too much," feeling kicks in after your stomach contains between 16 and 50 fluid ounces. But the stomach can stretch out over time, especially if you practice over-eating and ignore your body's signals. For example, the 2017 Nathan's Hot Dog Eating champ downed 72 hot dogs. That's about two gallons worth of food in his stomach. Ultimately, the **10**_____ has the final say in how much you can eat.

Increased Exposure to Sunlight May Be Good for Some People (32)

Directions: Fill in a word that best completes each sentence.

A study by scientists at the U.S. Department of Energy's Brookhaven National Laboratory and colleagues in Norway suggest that the benefits of moderately increased exposure to sunlight, such as the production of vitamin D, may outweigh the risk of developing skin cancer. (**1**_____ D protects against the lethal effects of many forms of cancer).

"We know that solar radiation is the leading cause of skin cancer," said author Richard Setlow, of Brookhaven. Setlow's group was the first to establish that ultraviolet A (UVA) radiation and visible light are the primary causes of malignant melanoma, the deadliest form of skin cancer. He and his colleagues emphasize that people need to protect themselves from the harmful effects of sun exposure.

Solar radiation is the main source of vitamin D in humans. In the presence of sunlight, the body converts certain precursor chemicals to active vitamin **2**_____. "Since vitamin D has been shown to play a protective role in a number of internal cancers and possibly a range of other diseases, it is important to study the relative risks to determine whether advice to avoid sun **3**_____ may be causing more harm than good in some populations." The concern is great in populations from northern latitudes where **4**_____ exposure is limited.

In the current study, Setlow and his colleagues calculated the relative production of vitamin D via sunlight as a function of latitude, or distance from the equator. The scientists also examined the incidence of and survival rates for various forms of cancer by latitude. According to the calculations, people living in Australia (just below the equator) produce 3.4 times more vitamin D because of sun exposure than people in the United Kingdom, and 4.8 times more than people in Scandinavia. "There is a clear north-south gradient in vitamin D production," Setlow says, "with people in the north producing less than people nearer the **5**_____."

The scientists also found that the incidence rates of internal cancers such as cancers of the colon, lung, breast, and prostate increased from north to south. However, when the scientists examined the survival rates for these cancers, they found that people from the southern latitudes were less likely to die from those internal **6**_____ than people in the **7**_____. Previous research indicates that good vitamin-D status is advantageous when combined with standard cancer therapies. Says Setlow, "The current data provide a further indication of the beneficial role of sun-induced vitamin D for cancer prognosis."

"As far as skin cancer goes, we need to be most worried about melanoma, a serious disease with significant mortality," Setlow says. **8**_____ is triggered by UVA (long UV wavelengths) and visible light. Vitamin-D production in the body, on the other hand, is triggered by UVB (short **9**_____ wavelengths at the earth's surface). "Perhaps we should redesign sunscreens so they don't screen out as much UVB while still protecting us from melanoma-inducing **10**_____ and visible light," Setlow says.

Veterinary Surgeons Perform First-Known Brain Surgery to Treat Hydrocephalus in Fur Seal (33)

Directions: Fill in a word that best completes each sentence.

A neurosurgical team at Cummings School of Veterinary Medicine has successfully performed the first-of-its-kind brain surgery on a Northern fur seal named Ziggy Star. The **1**_____ was an attempt to address her worsening a neurologic condition called hydrocephalus. The **2**_____, common in cats and dogs, has not been well documented in pinnipeds (seals, sea lions, and walrus).

"The ability to successfully complete this procedure with many unknown factors is due in large part to the collaboration among colleagues at Cummings and Mystic," said Cummings lead neurosurgeon, Ane Uriarte, DVM. "The combined expertise and skills from all our areas of specialty -- from neurosurgery to anesthesia and zoological medicine -- was critical to this success."

Mystic Aquarium took in Ziggy approximately four years ago after she was found stranded on the California coast and deemed non-releasable by the federal government. At the time, she had an MRI that showed a **3**_____ condition. She received treatment, but her symptoms continued to progress, with the seizures emerging more recently.

Said Uriarte, "After discussion with Mystic's veterinary team, we determined the best option to prevent further deterioration of the brain and to improve Ziggy's symptoms was to surgically place a shunt to drain the excess fluid, relieving some of the pressure on the brain." While this surgical procedure could not reverse damage caused to the **4**_____ by excess fluid, if successful, it could stop the progression of Ziggy's condition, improving her quality of life, level of responsiveness and mobility.

The team present on **5**_____ day included veterinary anesthesiologists, neurosurgeons and zoological medicine specialists from Cummings Veterinary Medical Center, as well as zoological medicine specialists from Mystic **6**_____ who serve as Ziggy's primary veterinarians. The procedure involved placing a **7**_____ -- a narrow tube -- through the skull and into the brain. This tube was then positioned underneath the skin through the neck and passed down to Ziggy's abdomen. A valve controls the flow of excess cerebral spinal fluid from the brain to the **8**_____, where it is absorbed by the body.

Ziggy was transferred back to Mystic Aquarium once she was in stable condition. By December 2017, **9**_____ was living in an off-exhibit habitat at **10**_____ Aquarium, where she was monitored through her recovery and rehabilitation.

"We continue to monitor Ziggy very closely," said Jen Flower, DVM, MS, Diplomate of the American College of Zoological Medicine, Chief Clinical Veterinarian at Mystic Aquarium. "She is showing marked progress daily; eating a full diet; moving well within her habitat and showing normal swim patterns. No additional seizures have been noted post-operatively."

Why Are Suicide Rates Rising? (34)

Directions: Fill in a word that best completes each sentence.

Suicide is becoming more common. The Centers for Disease Control and Prevention (CDC) say that the rates of death by **1**_____ in the US have risen by 25 percent in the last couple decades. Though the reasons for that increase are not completely clear. Experts have previously pointed to an increased sense of isolation among Americans, as well economic factors, and a rise in mental illness. Others point to the rise of technology, which has replaced important face-to-face interactions as a factor in the suicide increase.

These explanations are speculative, however. It's hard to make broad statements about suicide, said Dr. Katalin Szanto, who has published widely on suicide prevention. For instance, suicide is now the second leading cause of death for people ages 15 to 24 in the U.S., yet many researchers think aging Baby Boomers will be especially vulnerable to suicide in the coming years. Another paradox is that suicide in the U.S. is often connected to other forms of violence, such as bullying, sexual violence or child abuse, yet rates of those forms of **2**_____ have not increased in the past twenty **3**_____.

Szanto says that if people can find help the first time that they consider suicide, they are more likely to recover and never make another attempt. Yet, people who have tried once are much more likely to **4**_____ again. A 10-year study at the Henry Ford Hospital in Detroit, found that it is possible to stop individual suicides. Doctors and therapists employed several interventions that led to an 80-percent drop in suicide rates. One such method involved asking depressed patients how they envisioned dying. Doctors then created roadblocks to enacting that vision. For example, they asked **5**_____ to remove firearms from their houses and then followed up to see if they did so.

However, knowing what drives suicide and having improved treatment options won't help if people don't reach out for help when they are at the their most hopeless, said Susan Lindau, a practicing therapist who specializes in suicide. One finding in the new CDC report is that more than half of deaths happened among people who had not been diagnosed with mental **6**_____. Said Szanto, "We have this big problem, especially among men, that they have undiagnosed and obviously untreated **7**_____ health conditions. Often the manifestation of depression is different in **8**_____ than in women. We tend to be a little bit better in our clinical assessments to diagnose 'typical' depression in **9**_____."

Lindau points to the problem of stigma around depression and mental illness. "It is very brave to be able to say, 'I feel horrible and I need to reach out.' Because you are revealing your vulnerabilities. Our culture does not really respect vulnerability." She added that people need to understand that mental illness is chronic like diabetes or multiple sclerosis. For many people in crisis, it is important that they reach out to family or friends and get through that moment they are considering **10**_____. The pain won't disappear, but they have better odds of coming through the other side and moving toward treatment and recovery, Lindau said.

You Are Disgusting in So Many Ways (35)

Directions: Fill in a word that best completes each sentence.

Humans evolved the ability to feel revolted by awful sights, sounds, smells, and tastes for good reason: feeling disgusted causes us to avoid things that are potentially dangerous. A new study, published in the Royal Society's Philosophical Transactions B, originated from the idea that distaste acts as an "intuitive microbiologist" that tracks the sources of infection in our environment and motivates humans to **1**_____ those things. "You can think of it as a behavioral arm of the immune system," says study coauthor Mícheál de Barra, of Brunel University in London. The immune **2**_____ protects the body from disease. Avoidance is a **3**_____ arm offering protection, as well. That's not necessarily a new idea, "but so far, there has been little attempt to link the sources of disgust to the sources of infectious disease in a comprehensive way." This study makes a step toward linking **4**_____ of disgust with disease.

Barra and his colleague, Val Curtis, of the London School of Hygiene and Tropical Medicine, developed a method for understanding **5**_____ based on illnesses, by asking participants to rate their response to more than 70 different mortifying scenarios, like encountering hairless cats, touching scabby fingers, learning a friend tried to have sex with pieces of fruit, and seeing pus from genital sores. The result is a new framework that **6**_____ and Barra hope will resolve some questions about the kinds of things people find disgusting, and why.

By coding the research subjects' responses, Curtis and **7**_____ found six categories of disgust: poor hygiene (snotty tissues, body odor, a dirty apartment bathroom), animals and pests (cockroaches, rats, infestations), sexual behavior (prostitution, promiscuity), irregular or strange appearances (obesity, disfigured faces, amputated body parts, poverty, wheezy breathing), lesions or visible signs of infection, and rotting or decaying food.

Why humans feel revulsion to certain things is understandable, but the origins behind how and why humans find particular things repulsive is less clear. Disgust, says Barra,"is simultaneously emotional, biological and cultural—certainly a combination of learning and genes." Seeing smallpox sores will cause disgust in almost anyone, because you instinctively know it's a sign of **8**_____. But meanwhile, one person's old flimsy couch tossed on the street really can be another person's treasured new centerpiece of home decor. "The idea of eating from a clean dog bowl is disgusting because of a learned association," says Barra. Genes might decide what kills us and what doesn't, but it's through our interactions with the environment and with other people that we learn how to calibrate and adjust to our surroundings. While the six **9**_____ might broadly encompass most disgusting things, there will variability depending on a person's previous interactions with their **10**_____ and life experiences.

Answer to Young People's Persistent Sleep Problems (36)

Directions: Fill in a word that best completes each sentence.

A research project involving James Cook University and the University of Queensland indicates high rates of sleep problems continuing through teenage years and into early adulthood -- but also suggests a natural remedy.

Dr. Fatima from JCU was associated with a study that tracked more than 3600 people from the age of 14 until they were 21. "Just over a quarter of the 14-year-olds reported sleep problems, with more than 40 percent of those still having **1**_____ problems at 21," said Dr. Fatima. She said the causes of sleep **2**_____ were different at different ages.

"Maternal factors, such as drug abuse, smoking, depression and anxiety among mothers are the most significant predictors of adolescent sleep problems in their **3**_____, at 14-years-old. For all people studied, being female, having experienced early puberty, and being a smoker were the most significant predictors of sleep problems at 21 years." She said adolescent depression or anxiety were linking factors for sleep problems between the two **4**_____.

"It's a vicious circle. Depression and anxiety are well-established risk **5**_____ for sleep problems and people with sleep problems are often anxious or depressed," she said.

Dr. Fatima said that as well as the traditional factors, excessive use of electronic media is emerging as another significant **6**_____. "In children and adolescents, it's found to be strongly associated with later bedtime and shorter sleep duration, increasing the risk of developing **7**_____ disturbances," she said.

Dr. Fatima said the study was worrisome as it revealed a high incidence of persistent sleep problems and possible concurrent health **8**_____ among young people -- but it also strongly suggested an answer to the problem. Researchers found that an active lifestyle can decrease future incidence and progression of sleep problems in young subjects. Dr. Fatima said, "Early exercise intervention with adolescents might provide a good opportunity to prevent their sleep problems persisting into later **9**_____."

The next study being considered would look at what factors lead to young adults' sleep problems continuing as they grow **10**_____ and how that might be prevented.

Reproduction & Development

New Gene May Explain Why Some Men More Likely to Have Sons (1)

Directions: Fill in a word that best completes each sentence.

A Newcastle University study involving thousands of families is helping prospective parents work out whether they are likely to have sons or daughters. The work was done by Corry Gellatly, a research scientist at the university. The study examined 927 family trees containing information on 556,387 people from North America and Europe going back to 1600. "The **1**_____ tree study showed that whether you're likely to have a boy or a girl is inherited. We now know that men are more likely to have sons if they have more brothers but are more likely to have daughters if they have more **2**_____. However, in women, you just can't predict it," Mr Gellatly explains.

Baby gender is determined by whether a male's sperm is carrying an X or Y chromosome. An X chromosome combines with the mother's X **3**_____ to make a baby girl (XX) and a Y chromosome combines with the mother's X chromosome to make a boy (XY). All chromosomes contain genes.

Genes consist of two parts, known as alleles. One allele is inherited from each parent. In his paper, Mr Gellatly provides new information that shows that it is likely that men, instead, carry different types of alleles, which result in three possible combinations in a (currently unidentified) gene which controls the ratio of X and Y sperm: Men with the first combination (mm) produce more Y sperm and have more sons; men with the second combination (mf) produce an equal number of X and Y **4**_____ and have an equal number of sons and daughters; men with the third combination (ff) produce more X sperm and have more daughters.

Mr Gellatly hypothesizes that the currently unidentified **5**_____ is passed on from both parents. He thinks this gene causes some men to have more sons and some to have more **6**_____. This may explain why we see the number of men and women roughly balanced in a population. If there are too many males in the population, for example, females will more easily find a mate, so men who have more daughters will pass on more of their genes, causing more females to be born in later generations.

This may also be the reason why there are more male births after wars. The odds are in favor of men with more sons having at least one **7**_____ return from the war. Those sons are then more likely to father boys themselves because they inherited that tendency from their fathers. In contrast, **8**_____ with more daughters may have lost their only sons in the **9**_____ and those sons would have been more likely to father girls. After wars there is often a "boy-baby boom".

In most countries, for as long as records have been kept, more boys than girls have been born. In the UK and US, for example, there are currently about 105 males born for every 100 females. It is well-documented that more males die in childhood and before they are old enough to have children. So in the same way that the gene may cause more **10**_____ to be born after wars, it may also cause more boys to be born each year.

Our Mothers' Exposure to BPA Might Lead us to Binge Eat as Adults (2)

Directions: Fill in a word that best completes each sentence.

Bisphenol A (BPA) has been removed from baby bottles, sippy cups, and infant formula packaging because it has been found to leak out of plastics which are made from it. There is evidence that high doses BPA may disrupt developing brain and endocrine systems. BPA is, however, still present in many plastic food containers.

Exposure to BPA before birth changes the way the brain deals with signals of "hungry" or "full" later in **1**_____, according to a study in *Endocrinology*. Researchers found that when they fed pregnant mice low doses of BPA they gave birth to pups with underdeveloped brain circuitry in areas related to satiety, or how much they want to **2**_____. During **3**_____ development, the hormone leptin surges in the brain and helps develop the brain's sense of satiety (fullness). If this developmental **4**_____ surge does not happen on time, it cannot be restored in adults.

Leptin has a similar function in humans. Children that have genetic defects in their **5**_____ signaling system are obese (extremely fat). We can't say that BPA increases body **6**_____. BPA changes the ways the **7**_____ understands metabolic signals from leptin, which can lead to weight **8**_____.

Pregnant and lactating women (those producing milk for nursing babies) need to be cautious. There are many plastics in society, but there are simple things they can do to reduce exposure to BPA. For example, heating **9**_____ can release the chemicals, so keeping Tupperware out of the dishwasher and microwave can help. Frozen dinners should be heated without the plastic container, either on a paper or glass plate.

BPA can still get into **10**_____ through the placenta and breast milk of the mother, so the government should play a roll, as well. The FDA and EPA need to work with industry to evaluate the impact of artificial chemicals on human health to examine how chemicals might influence the brain and change the way the body functions later in life.

Scientists are Puzzling Out How Butterflies Assemble Their Brightly Colored Scales (3)

Directions: Fill in a word that best completes each sentence.

From the danger-sign orange of Monarch butterflies, to the regal blues of the Blue Morpho, butterflies come in a rainbow of **1**_____. The insects display those incredible hues thanks to scales on their wings. Those scales are made up of crystals, which are made up of a sugar molecule called chitin (the same stuff that makes up insect exoskeletons and mushrooms). The tiny crystals on butterflies' scales are called gyroids. They're of interest to biologists, but also to materials scientists because the butterflies' gyroids are much tinier and more precise than anything made by humans today.

A new paper in Science Advances explores how those crystals develop on a butterfly wing while it's metamorphosing from a caterpillar into its beautiful final form. The lead author, Bodo Wilts, explains that, "Butterfly pupation normally lasts about 2 weeks. The scales grow in about 2-3 days from a first identifiable shape to the full scale, depending on the species of **2**_____ and the conditions of rearing."

Researchers can't yet put a camera inside a cocoon to watch the process in real time, although other labs are working solving that problem. Current **3**_____ aren't small or sensitive enough to watch things like butterfly scales forming and the wet interior of **4**_____ would harm sensitive electronic **5**_____.

Instead of trying to watch the scales and gyroids develop in real-time video, Wilts and colleagues looked at the mature scales of the small Hairstreak butterfly *Thecla opisena* from Mexico under a very strong microscope. That let them see details of the wing's structure. Wilts expected to find the 'usual' pattern of green scales observed in many **6**_____, but instead found that each wing **7**_____ carried extremely small crystals that were (surprisingly) not interconnected.

In this butterfly, the gyroids were arranged in lines from small to large along a single scale, giving the **8**_____ its shine and pigment, and giving researchers a glimpse at how they might have formed. They say that instead of growing in a pre-set template within the **9**_____, the shape of the structures seems to grow in a constantly changing manner over time.

This study only looked at one of the over 140,000 species of **10**_____, but there is a good chance that this finding could tell researchers more about how other distinctive wing patterns form. In fact, scientists believe that the growing mechanism found in *Thecla opisena* is generalizable to most wing scales.

Learning and unravelling the principles behind these complex folding patterns on the molecular level will be very important because these tiny structures could tell researchers more about other organisms and structures, like mitochondria in cells, which have similarly complex patterns.

Scientists Discover the Embryonic Origins of the Phallus (Genital Organs) (4)

Directions: Fill in a word that best completes each sentence.

Tracking which embryonic cells become what body part is an important part of figuring out the basic rules for making life. For example, scientists know, in detail, what **1**_____ turn into limbs and eyes. But, until recently, scientists did not understand which embryonic cells become the genitals. Biologists have now published the first reports of cell-fate-tracking for phalluses in snake, chick, and mouse embryos. Phallus is the term for both clitorises and penises.

Considering the similarities between mouse and human embryo development, the mouse results are likely to apply **2**_____ phallus development, as well. Patrick Tschopp, a geneticist at Harvard University says, "We can be fairly confident that in humans, it's pretty similar."

In mice, a structure in the embryo called the cloaca sends chemical signals to another structure, called the tail- bud. Some of those tail-**3**_____ cells form the genital tubercle, which eventually becomes a penis or clitoris. Scientists knew about the genital **4**_____ before, but never knew where it came from. Meanwhile, the rest of the **5**_____-bud cells become tails, in mice at least. Human embryos' tail-buds eventually disappear.

Interestingly, the situation is a bit different in snakes and chicks. Snakes' phalluses originate from limb-buds which, in snakes, are evolutionary leftovers, like human tail-buds. Chicks' **6**_____ draw from both **7**_____-buds and tail-buds, Tschopp's team found. All three species' phallic development depends on signals from the cloaca, which eventually forms the urethra and anus in mammals.

Knowing the origin of phalluses further back in an embryo's history helps scientists figure out what genes drive phallus development. Scientists can look in those progenitor (beginning) cells to see what genes are turned on. That, in turn, may pinpoint which **8**_____ fail to function properly in human babies with genital birth defects, says Martin Cohn, a biologist at the University of Florida who worked on the research team. **9**_____ birth defects can be fairly common, but their causes are usually mysterious.

In addition, knowing how phalluses grow in different species offers a clue into how phalluses evolved in general. Phalluses are among nature's fastest-evolving body parts, as Cohn has found in previous research. Consider the differences between lizard and snake penises and mammalian ones. Male lizards and **10**_____ each have a pair of sex organs called hemipenes, which are not enclosed. Instead, they act like slides for sperm. Turtles, birds, and crocodile-like animals also have slides instead of tubes.

It is still unclear why the animals of the world have evolved such a diversity of phalluses.

Cells Driving Gecko's Ability to Re-grow its Tail Identified (5)

Directions: Fill in a word that best completes each sentence.

Many lizards can detach a portion of their tail to avoid a predator and then regenerate a new one. Unlike mammals, the lizard **1**_____ includes a spinal cord. Prof. Matthew Vickaryous found that the spinal cord of the tail contained many stem cells and proteins known to support stem cell growth.

"We knew the gecko's spinal cord could **2**_____, but we didn't know which cells were playing a key role," said Vickaryous, lead author of the study recently published in the *Journal of Comparative Neurology*. Vickaryous has been studying **3**_____ in the hopes of finding a way to get human spinal **4**_____ to repair themselves after injury like those in the gecko.

Geckos can re-grow a new tail within 30 days -- faster than any other type of **5**_____. In the wild, they detach their tails when grabbed by a predator. The severed tail continues to wiggle, distracting the **6**_____ long enough for the reptile to escape.

In the lab, Vickaryous simulates this by pinching the gecko's tail causing the tail to drop. Once detached, the site of the tail loss begins to repair itself, eventually leading to new tissue formation and a new spinal cord. For this study, the biomedical sciences professor, along with PhD student Emily Gilbert, investigated what happens at the cellular level before and after detachment.

They discovered that the spinal cord houses a special type of stem cell known as the radial glia. These **7**_____ cells are normally at **8**_____. However, when the tail comes off everything temporarily changes. The cells make different proteins and begin multiplying in response to the injury. Ultimately, they make a brand-new spinal cord. Once the injury is healed and the spinal cord is restored, the cells return to a resting state.

Humans, on the other hand, respond to a spinal cord injury by making scar tissue rather than new tissue. The **9**_____ tissue seals the wound quickly, but sealing the injury prevents regeneration. It's a quick fix, but in the long term it's a problem. The scarring may play a role in why humans have a limited ability to repair their spinal cords. Humans are missing the key cells types required

This study is part of a series of investigations into the regenerative abilities of the gecko's central nervous system. The next step is to examine how the gecko is able to make new brain cells. Geckos can regenerate many tissues throughout their bodies, making them ideal models for studying wound healing and **10**_____ re-development. We can learn a lot from them.

Heteropaternal Superfecundation: Who's Your Daddy? (6)

Directions: Fill in a word that best completes each sentence.

An extremely rare occurrence in humans, heteropaternal superfecundation, would only be discovered if a family is suspicious that a pair of fraternal twins were sired by two different fathers. One study estimated the rarity of this condition to be one in 400 (0.25%) twin births in the US. Another study reported that, among non-identical twins whose parents had been involved in paternity suits, the frequency was 2.4%.

For heteropaternal **1**_____ to occur, a female would have to ovulate twice within a few days (a rare, but not unheard-of, phenomenon) and then have sex with two different men during that time. If each coital (sex) act yielded a zygote, each **2**_____ would need to then mature into a fetus and be born. The live births would result in fraternal "twins" having different fathers. That condition would be **3**_____ superfecundation.

Most of the time, women ovulate a single egg per menstrual cycle. One egg, one shot at fertilization, with ovulation stopping as soon as pregnancy occurs or, alternatively, without fertilization, till the next menstrual **4**_____. In extra-ordinary circumstances, if a woman ovulates more than one **5**_____ during a single **6**_____ cycle, she is termed fecund. The term heteropaternal superfecundation is a way to denote that a super-fertile (fecund) woman is carrying two or more babies fathered by two or more distinct **7**_____.

Though heteropaternal superfecundation is rare in humans, it is common among cats. Female cats, also called "queens," usually give birth to litters of 3 to 5 kittens, but as many as 19 have been born at once. Kittens from the same litter can look completely different. This is because **8**_____ can be impregnated by more than one male, or tomcat, during a single ovulation period, due to heteropaternal superfecundation.

These kittens are like fraternal twins, yet they are genetically different. They occupy the uterus together. Yet, each kitten could be sired by a different tomcat. Heteropaternal superfecundation has been reported in other mammals, including dogs and cows, due to the number of eggs ovulated per cycle. Though possible in humans, it rare, since human females typically ovulate a single egg per cycle.

In cats, heteropaternal superfecundation can be evolutionarily advantageous: the more **9**_____ a queen mates with, the more likely she is to have kittens. Unlike the tomcat, the queen's genetic makeup is in every one of the **10**_____.

To curb the unwanted overpopulation of animals, such as cats, spaying or neutering is an encouraged practice.

Armadillo, Hedgehog and Rabbit Genes Reveal How Pregnancy Evolved (7)

Directions: Fill in a word that best completes each sentence.

There are about 4,000 species of placental mammals, which include *Homo sapiens*. Placental **1**_____ are unique in their ability to nurture a fetus within their bodies for extended periods of time. This adaptation permits the slow development of big brains.

A new study of gene expression in early pregnancy suggests that placental mammals evolved the ability to turn the body's inflammatory response to the embryo implanting in the uterus into an advantage. "Implantation looks like inflammation." says Arun Chavan, an evolutionary biologist at Yale University. He presented the findings at the Society for Integrative and Comparative Biology meeting in California. "We did this study to learn how it became an implantation process instead."

Evolutionary biologists think that all ancient mammals laid eggs, as platypuses do today. Marsupials (like opossums and kangaroos) evolved later. Marsupials' fetuses hatch from a shelled egg within the mother and are then expelled from her body shortly after. Chavan and his colleagues showed that a group of inflammatory genes turn on as an opossum fetus leaves the **2**_____ and clings to the uterine lining.

In placental mammals, the embryo does not simply cling to the uterus. Rather, it destroys the **3**_____ lining as it invades the tissue, which triggers inflammatory proteins. During infection and injury, inflammatory **4**_____ normally fight invaders and repair wounds. Some of these proteins could harm the embryo, but inflammation may be necessary to keep it alive because inflammation causes the growth of new blood vessels, which provide the developing embryo the oxygen and nutrients it needs.

To find out how **5**_____ mammals can withstand the **6**_____ proteins in these early days, Chavan analyzed the inflammatory response in three species: the rabbit, armadillo, and hedgehog. He found that a normally active protein, interleukin-17, was inactive in placental mammals. When interleukin-17 is active, it signals to white blood cells to kill invaders through digestion or enzyme action. Chavan believes that the interleukin-**7**_____ would damage the embryo if it were active in the early pregnancy of placental mammals.

Chavan found that cells lining the **8**_____ of placental mammals suppress the production of **9**_____-17 early in pregnancy. Placental mammals seem to have a fine-tuned inflammation over the course of pregnancy that cycles up and down. The bodies of placental mammals have evolved to keep the aspects of the inflammatory response that are favorable to the fetus and stop the destructive parts of the response. Researchers are currently unsure how the inflammatory **10**_____ subsides later in pregnancy.

Doctors want to reduce miscarriages and improve implantation rates for women undergoing in vitro fertilization. Although inflammation appears to be necessary during implantation, it's a leading cause of miscarriage. Doctors are searching for ways to help switch from pro- to anti-inflammatory states.

Can We Stay Young Forever, Or Even Recapture Lost Youth? (8)

Directions: Fill in a word that best completes each sentence.

In 1962 Leonard Hayflick discovered that human cells have a limited replicative lifespan. This "Hayflick limit" of cellular lifespan is directly related to the number of unique DNA repeats found at the ends chromosomes. These DNA **1**_____ are part of the protective end-capping structures, termed "telomeres," which safeguard the chromosomes from DNA rearrangements that destabilize the genome.

Each time the cell divides, the telomeres shrink and will eventually fail to secure the chromosome ends. This reduction of telomere length functions as a "molecular clock" that counts down to the end of cell growth. This phenomenon is associated with the aging process, weakness, illness, and organ failure.

The enzyme telomerase counteracts this **2**_____ shrinking process. Telomerase offsets cellular aging by adding back lost DNA repeats to add time onto the molecular **3**_____, extending the life of the cell. **4**_____ lengthens telomeres by repeatedly synthesizing short DNA repeats of six nucleotides -- the building blocks of DNA -- with the sequence "GGTTAG" onto chromosome ends. However, the activity of the telomerase is insufficient to completely restore the telomeres, nor stop aging.

The gradual shrinking of telomeres decreases replication of adult stem cells, the cells that restore damaged tissues and replenish aging organs in our bodies. Telomerase slows the **5**_____ clock, it does not immortalize these cells. Better understanding of the telomerase enzyme could help reverse telomere shortening, cellular aging, extend lifespan, and improve the health of elderly individuals.

Research from the laboratory of Julian Chen and his colleagues at Arizona State University recently uncovered a step in the telomerase catalytic cycle that limits the ability of telomerase to synthesize repeats on **6**_____ ends. "Telomerase has a built-in braking system to ensure precise synthesis of correct telomeric DNA repeats. This brake, however, also limits the activity of the telomerase enzyme," said Professor Chen.

This brake is a pause signal for the enzyme to stop DNA synthesis at the end of the sequence 'GGTTAG'. When telomerase restarts, this pause signal is still active and limits DNA **7**_____. If scientists can specifically target the pause **8**_____ that limits DNA repeat synthesis, the telomerase can be supercharged to better work against telomere length reduction, with the potential to rejuvenate aging human adult **9**_____ cells. This sort of treatment could bring youth to aging cells and treat diseases which cause organ deterioration.

Scientists need to be careful, however, because it's not just stem cells that use telomerase. Cancer cells also use telomerase for destructive growth. Somatic cells constitute most of the cells in the human body and they have little telomerase, which reduces their risk of cancer development. Drugs that increase telomerase activity in all cell types are not desired, because they would increase **10**_____ risk in somatic cells. Scientists need to target the increased telomerase activity only in adult stem cells.

How Mum's Immune Cells Help Her Baby Grow (9)

Directions: Fill in a word that best completes each sentence.

Scientists have found that immune system cells, which are normally involved in defending against infection, also provide growth hormones to developing embryos. These findings may help prevent some miscarriages.

A week or two after being conceived, the human embryo is just a tiny ball of cells. After this tiny ball of **1**_____ travels through the oviduct, it must perform implantation. This is when the embryo attaches to the wall of the mother's uterus and embeds itself into the tissue. The cells on the outside of the **2**_____ then start to form the placenta, which is the food and waste transport system linking the mother with her developing offspring.

After **3**_____, the mother deploys large numbers of a special kind of white blood cell to the placenta. These are called "Natural Killer", or "NK" cells, which are normally responsible for killing virus-infected cells in the body. In pregnancy, though, they have a very different role. Scientists have known for some time that these cells are important for the healthy development of the **4**_____ and its blood vessels, but the full picture of their contribution wasn't appreciated.

The findings of a group of researchers at the University of Science and Technology of China were published in the journal *Immunity*. The scientists took uterine NK cells from pregnant **5**_____ and studied the pattern of genes that are active in the cells. A subset of the NK cells was producing growth-promoting factors, known to trigger the growth of bone, cartilage and blood vessels in an embryo.

The researchers then compared healthy pregnancies with those that ended in miscarriage. **6**_____ ending in miscarriages, they found, often had fewer of these growth-promoting **7**_____ cells, suggesting that NK **8**_____ contribute to embryo development through the expression of these growth factors.

To test this idea, the team then used genetically modified mice lacking growth-promoting NK cells in the uterus. Embryos of these mice, they found, tended to be much smaller than those of normal **9**_____, and their bones didn't form properly. If growth-promoting NK cells were injected into the genetically-modified mice though, their embryos then developed normally.

These findings could have important clinical implications. If a woman who wants to have a child doesn't have enough growth-promoting NK cells in her uterus, it's possible that her pregnancy will end in **10**_____. But if these cells can be grown in the lab, the team speculate, they could be added to her uterus to help the embryo develop normally.

An Old Drug for Alcoholism Finds New Life as Cancer Treatment (10)

Directions: Fill in a word that best completes each sentence.

In 1971, at the age of 38, a patient's breast cancer had metastasized (spread) to her bones. This is typically fatal, but the patient continued living. The patient then became alcoholic so her doctors stopped cancer treatment in favor of giving her a drug that discouraged drinking. She died at age 48 from a drunken fall. Her autopsy revealed that her bone tumors had melted away, leaving only a few cancer cells in her **1**_____ marrow. This report, and other studies, have suggested that the drug disulfiram (also known as Antabuse), which makes people feel sick from drinking small amounts of **2**_____, might also be a cancer fighter.

Scientists have known since the 1970s that disulfiram killed cancer cells and slowed tumor growth. Still, the drug hadn't gotten much attention for treating **3**_____, because scientists disagreed about how it worked. Researchers have now found that disulfiram blocks a molecule that helps get rid of cell waste.

In the new study, a group of Danish-Czech and U.S. researchers went through Denmark's registry of 240,000 cancer cases and examined the medications each patient took. Of the more than 3000 patients taking Antabuse, the cancer death rate was lower for the 1177 who stayed on the drug compared with those who stopped taking it. The **4**_____ had benefits for prostate, breast, and colon cancer.

The researchers also confirmed that disulfiram slows the growth of breast cancer tumors in mice, particularly if combined with a copper supplement, which was already known to enhance its effects. When disulfiram broke down within the bodies of the mice, one of its components joined with copper to cause a protein buildup within the cancer cells, causing them to die, explained cancer biologist Jiri Bartek of the Danish Cancer Society Research Center in Copenhagen, a co-leader of the study. Although other cancer drugs interfere with the same **5**_____ process in cancer cells, disulfiram is more specific, which could explain why it is so effective.

Even with these promising results, **6**_____ probably "is not a cure" for most cancer patients, cautions cancer biologist Thomas Helleday of the Karolinska Institute in Stockholm. However, the drug could help extend the lives of **7**_____ with metastatic (spreading) cancer—it's already shown evidence of doing so when combined with chemotherapy in a small lung cancer trial. Bartek and collaborators are now launching clinical trials to test a disulfiram-copper combo as a treatment for metastatic (**8**_____) breast and colon cancers and for glioblastoma, a type of brain cancer.

Finding a new use for an approved **9**_____ is appealing because the compound has already passed safety testing. However, big pharmaceutical companies probably won't be interested in developing disulfiram for cancer because there's no patent protection on the drug, Bartek says. Still, if clinical **10**_____ provide convincing evidence oncologists (cancer doctors) could prescribe it anyway as an inexpensive treatment.

Self-Fertilizing Fish Have Surprising Amount of Genetic Diversity (11)

Directions: Fill in a word that best completes each sentence.

The mangrove killifish flourishes in both freshwater and water twice as salty as the ocean. It can live up to two months on land, breathing through its skin, before returning to the **1**_____ with a series of spectacular 180-degree flips. It is also unusual because it is a rare vertebrate that fertilizes itself.

Self-fertilization is beneficial when food is plentiful, and the climate is good because an organism can reproduce. But, because a self-**2**_____ animal replicates only its own DNA, it is deprived of a diverse genetic toolkit that can be handy if the environment suddenly changes. **3**_____-fertilization also can increase the odds of a dangerous mutation becoming common in a population.

Luana Lins, of the Washington State University, and her colleagues, sequenced the genome of the **4**_____, also known as *Kryptolebias marmoratus*, or "Kmar," they learned that there was great genetic diversity across the species. The researcher's findings are published in the journal, *Genome*.

When individuals mate, their chromosomes line up. Each parent's genes have different nucleotides, or building blocks, at corresponding locations of the DNA. These different **5**_____ are called heterozygous. Offspring have different traits from each parent since each provided unique genes.

But when a creature fertilizes its own egg with its own sperm, its nucleotides are more likely to match up uniformly, except for the occasional mutation or recombination. Their paired **6**_____ are called homozygous and, like clones, one generation is pretty much the same as the next.

In looking at killifish DNA from various lineages, one might expect to see a certain uniformity among them. Yet Lins and colleagues at WSU, Stanford University and the University of Alabama didn't. They found more areas that were heterozygous than they expected. The researchers wondered how that was happening.

One explanation is that the **7**_____ have a lot more mutations than previously thought. Still, **8**_____ have a certain regularity, making this much diversity unlikely. Another explanation is that when these self-fertilizing, hermaphrodite fish are exposed to fish from different lineages, they are more likely to lay more eggs with them than with fish from the same **9**_____. In other words, when a stranger comes to town, the Kmar capitalize on this new pool of exotic DNA. They are drawn to the stranger, or they respond to the **10**_____ by producing extra eggs.

For all their limited brain capacity, the fish aren't exactly dumb. Lins has noticed that they learn when it is time to eat, rising to be fed. Still, spotting a new pool of DNA would be a remarkable sort of perception, for a fish or any other creature. "How they know how different the other individual is, we don't know," said Lins. "There are a lot of unknowns, and I think that's the fun of science."

Sex Reversal in Bearded Dragons Creates Females That Behave Like Males (12)

Directions: Fill in a word that best completes each sentence.

The large, striking lizard of the Australian desert known as the bearded dragon lays genetically male eggs that can develop into females if the nest becomes particularly warm. This creates lizards that are reproductively **1**_____ but retain the size and personality characteristics of **2**_____—a unique trait in the animal kingdom. The study documenting these findings has been published in the Proceedings of the Royal Society.

In most animal species, sex chromosomes will decide whether an individual is male or female. In other animal species, environmental conditions like the temperature of a nest will decide (as in the case of crocodilians.)

Sex determination in bearded **3**_____ is unique because both factors play a part. A mother dragon might lay a clutch of **4**_____ which contain genetically male individuals, but if the nest temperature tips above a certain level for whatever reason, the male sex chromosomes in those eggs will be overridden, and the little dragons inside will stop developing into males and switch to functionally reproductive females. This multifactorial system of sex determination hasn't yet been discovered in other animals.

Evolutionary biologists from the Universities of Sydney and Canberra in Australia filmed this third sex of dragons—these sex-**5**_____ females, as they called them—and measured how they reacted to different stimuli compared to their "normal" male and female colleagues. The researchers kept track of personality traits like boldness, activity level, and exploratory behavior.

Remarkably, the sex-reversed females were bolder, more assertive, and more aggressive than the other females, and even more so than the other "normal" males.

These novel traits of a behaviorally male, but functionally female dragon seem to create an individual that has a higher evolutionary fitness than regular dragons (that were not exposed to a **6**_____ drop while in the egg-state). Because of their increased assertiveness, the **7**_____-reversed females might have more success in mating.

Evolutionary biologists speculate that double-factor control over sex **8**_____ may be a way that **9**_____ dragons adapt to environmental changes. But, co-author and **10**_____ biologist Rick Shine points out that "On the other hand, it might generate an over-reaction to sudden changes in climate. 50/50 sex ratios are the safe evolutionary option – a population that produced 100 percent male or female offspring because of a sudden change in nest temperatures would be in real trouble."

Skin Cells Remember Previous Injury (13)

Directions: Fill in a word that best completes each sentence.

When tissues are damaged, they enter a state of inflammation which causes immune cells and blood to enter the injured area. Cell growth rates increase, and these combined immune responses enable rapid repairs to be made, restoring the tissue to a healthy, healed state.

Now scientists have discovered an additional element to this repair process: prior experience. Studying mice, Rockefeller University researcher Shruti Naik and her colleagues have found that, following an earlier injury, stem cells within the damaged patch of skin retain a memory of the earlier damage. This enables them to swing into action more rapidly if a later injury occurs in the same **1**_____.

In the team's study, published in *Nature*, **2**_____ received small injuries placed in patches of previously-harmed **3**_____. These injuries healed 2.5 times faster than similar **4**_____ placed elsewhere on the bodies of the mice, or in healthy never-before-injured control animals.

The researchers ruled out immune cells as a cause of the memory effect by repeating the experiments in mice unable to produce white blood cells that heal wounds. They found that damage to previously-injured skin still **5**_____ at twice the rate compared with skin that had not received a prior injury.

Skin stem cells appear to be able to have a genetic memory of earlier harm, which primes and sensitizes specific genes required to kickstart a rapid reaction if another **6**_____ happens later in the same location. This occurs when parts of the stem cell's DNA unwinds to expose specific **7**_____ to enable them to respond quickly to any inflammation within the cell.

Researchers found that more than 2000 locations in stem cell's **8**_____ remained open up to 180 days after a skin injury occurred. These exposed "primed" genes were often the first ones to be switched on in cells exposed to a new cut or other damage. These changes allow skin **9**_____ cells the ability to accelerate their response to future stressors.

Beneficial as that sounds, it could also cause problems. Genetic changes that get stem **10**_____ ready to act more rapidly are often also related to a greater risk of cancer, or the uncontrolled growth of cells. These same stem cell memory effects could also contribute to autoimmune skin diseases, like psoriasis, and allergic conditions like eczema. If so, this scientific discovery may provide exciting opportunities for developing new therapies to treat both diseases.

Scientific Inquiry

Behavior in High School Predicts Income and Occupational Success Later in Life (1)

Directions: Fill in a word that best completes each sentence.

Being a responsible student, maintaining an interest in school and having good reading and writing skills will not only help a teenager get good grades in high **1**_____ but could also be predictors of educational and occupational success decades later, regardless of IQ, parental socioeconomic status or other personality factors, according to research published by the *American Psychological Association*.

"Educational researchers, political scientists and economists are increasingly interested in the traits and skills that parents, teachers and schools should foster in children to enhance chances of success later in **2**_____," said lead author Marion Spengler, PhD, of the University of Tübingen. "Our research found that specific behaviors in **3**_____ school have long-lasting effects for one's **4**_____ life."

Spengler and her coauthors analyzed data collected by the American Institutes for Research from 346,660 U.S. high school students in 1960, along with follow-up data from 81,912 of those students 11 years later and 1,952 of them 50 years later. The initial high school phase measured a variety of student **5**_____ and attitudes as well as personality traits, cognitive abilities, parental socioeconomic status and demographic factors. The follow-up surveys measured overall educational attainment, income and occupational prestige.

Being a responsible student, showing an interest in school and having fewer problems with **6**_____ and writing were all significantly associated with greater educational **7**_____ and finding a more prestigious job both 11 years and 50 years after high school. These factors were also all associated with higher income at the 50-year mark. Most effects remained even when researchers controlled for **8**_____ socioeconomic status, **9**_____ ability and other broad personality **10**_____ such as conscientiousness.

While the findings weren't necessarily surprising, Spengler noted how reliably specific behaviors people showed in school were able to predict later success. Further analysis of the data suggested that much of the effect could be explained by overall educational achievement, according to Spengler. "Student characteristics and behaviors were rewarded in high school and led to higher educational attainment, which in turn was related to greater occupational prestige and income later in life," she said.

Caffeine's Sport Performance Advantage for Infrequent Tea and Coffee Drinkers (2)

Directions: Fill in a word that best completes each sentence.

Researchers Dr Brendan Egan and Mark Evans from the DCU School of Health and Human Performance examined the impact of caffeine, in the form of caffeinated chewing gum, on the performance of 18 male team sport athletes during a series of repeated sprints. The athletes undertook 10 repeated sprints under conditions with and without two sticks of the caffeinated gum, which is equivalent to two strong cups of coffee.

They found that the **1**_____ gum provided very little advantage to athletes whose bodies may have become desensitized to caffeine through a process called habituation, which occurs by having caffeine frequently, such as consuming coffee or sports drinks.

However, the athletes who had a low habitual caffeine consumption maintained their performance in repeated sprint tests after ingesting a caffeinated chewing **2**_____, while the performance of **3**_____ who consumed the caffeine equivalent of three or more cups of **4**_____ per day worsened over the course of the ten repeated **5**_____. This indicated that this second group did not benefit from caffeine as a performance aid.

Caffeine is regarded as one of the most popular performance enhancing supplements among athletes. The benefits of **6**_____ include improved muscle strength, mental alertness, as well as reducing the perception of effort during intense activity, therefore helping athletes to perform faster and longer.

The findings from the DCU-led study were published in the *International Journal of Sport Nutrition and Exercise Metabolism.* They recommended that athletes who consume caffeine on a regular basis should reduce their **7**_____ of coffee, **8**_____ drinks, or other products with **9**_____ in the lead-up to a big performance, if they want to receive the benefits of a caffeine supplement as a **10**_____ aid.

Cat Poop Parasites Don't Make You Psychotic (3)

Directions: Fill in a word that best completes each sentence.

Biologists have cited cat ownership as a risk factor for psychosis, schizophrenia, and other psychological problems. Studies have shown that cats can host the bacteria *Toxoplasma Gondii*, which has been linked to mental problems. The *T. Gondii* parasites get excreted in cat feces. Since cat owners often clean their cat's litter boxes, it follows that having a **1**_____ increases your risk of psychosis.

Except that it doesn't. This is a classic case of 'correlation doesn't imply causation'. Another example of correlation not implying **2**_____ would be if you compared the number of popsicles eaten to the number of drownings that occur. You would probably find a significant correlation between those two numbers: it seems like the more **3**_____ you eat, the more likely you are to drown. In fact, it's far more likely that there's a third, confounding factor: proximity to a body of water. People are more likely to eat popsicles when they're at the beach or by a pool, where drownings most often occur. Another example of **4**_____ not implying causation is people who have a lighter in their pockets being more likely to get lung cancer. Of course, this does not mean that lighters cause cancer, rather, it's because smokers often carry lighters and smoking causes lung **5**_____.

Correlation not implying causation may also be true of cat ownership: It's true a *T. Gondii* infection can lead to psychological problems because the parasite affects the muscles and brain. It's also true that *T. Gondii* reproduces and lives in cats and gets excreted in their feces. However, when psychiatrists at University College London observed over 4,500 children from birth to age 18, gathering data about their pet ownership, home lives, and health, the researchers couldn't find a link. Their study contradicts prior research indicating a correlation between cat ownership and **6**_____.

Prior studies found relationships between cat **7**_____ and mental health problems based on biased data. Many of the prior studies were retrospective, meaning they asked participants to look back on their lives and recall habits, living situations, and health issues. Recall like that is unreliable. Additionally, previous research used small sample sizes, adding to the reliability problem. Prior **8**_____ also did not control confounding factors, such as occupation, socio-economic level, other pet ownership, and over-crowding in the home. These confounding factors might, themselves, be contributors to psychosis.

It's also worth noting that even if your cat has a **9**_____ infection, it's unlikely that you'll get infected. As the Cornell Veterinary School points out, cats only shed the organism in their **10**_____ for a few days, which is a small amount of time in which a person could become infected. Plus, indoor cats who aren't fed red meat or don't hunt prey probably haven't caught the parasite to begin with. Notwithstanding any of these facts, cat owners should still be careful when cleaning litter boxes. It's just unlikely that the reason for taking care would be to avoid schizophrenia.

Dinosaur Parasites Trapped in 100-Million-Year-Old Amber Tell Blood-Sucking Story (4)

Directions: Fill in a word that best completes each sentence.

Researchers found a tick grasping a feather sealed inside a piece of 99 million-year-old Burmese amber. The discovery is amazing because fossils of parasitic, blood-feeding creatures attached to the remains of their host are very scarce. The new specimen is the oldest known, dating from the Cretaceous period (145-66 million years ago). The discovery is published in *Nature Communications.*

Enrique Peñalver from the Spanish Geological Survey and leading author of the work says that ticks are infamous **1**_____-feeding, parasitic organisms, which have a tremendous impact on the health of humans, livestock, pets, and even wildlife. Until now clear evidence of their role in deep time has been lacking.

Amber from the **2**_____ period provides a window into the world of the feathered dinosaurs, some of which evolved into modern-day birds. The feather found in amber with the grasping **3**_____ is similar in structure to modern-day bird feathers. It offers the first direct evidence of an early parasite-host relationship between ticks and **4**_____ dinosaurs.

Dr Ricardo Pérez-de la Fuente, a research fellow at Oxford University Museum of Natural History and one of the authors of the study explains that the fossil record tells us that feathers like the one we have studied were already present on a wide range of theropod dinosaurs, a group which included ground-running forms without the ability of flight, as well as bird-like dinosaurs capable of powered **5**_____,

Researchers can't be sure what kind of **6**_____ the tick was feeding on, but the mid-Cretaceous age of the Burmese **7**_____ confirms that the feather did not belong to a modern bird. Another kind of (extinct) tick, *Deinocroton draculi,* was also found sealed inside **8**_____ amber, with one specimen remarkably engorged with blood, increasing its volume approximately eight times over non-engorged forms. The specimen provided further indirect evidence that the ticks' host was a feathered **9**_____.

"The simultaneous entrapment of two external parasites -- the ticks -- is extraordinary and can be best explained if they had a nest-inhabiting ecology as some modern ticks do, living in the host's nest or in their own nest nearby," says Dr David Grimaldi of the American Museum of Natural History and an author of the work.

Together, these findings provide direct and indirect evidence that ticks have been parasitizing and sucking **10**_____ from dinosaurs within the evolutionary lineage leading to modern birds for almost 100 million years. While the birds were the only lineage of theropod dinosaurs to survive the mass extinction at the end of the Cretaceous 66 million years ago, the ticks did not just cling on for survival, they continued to thrive.

Does Dim Light Make Us Dumber? (5)

Directions: Fill in a word that best completes each sentence.

Spending too much time in dimly lit rooms and offices may change the brain's structure and hurt one's ability to remember and learn, indicates groundbreaking research by Michigan State University neuroscientists. The researchers studied the brains of Nile grass rats, which are diurnal (sleep at night).

Rodents exposed to dim light for four weeks lost about 30 percent capacity in the hippocampus, a critical brain region for learning and memory, and performed poorly on a spatial task (like moving around a maze) they had trained on previously. Conversely, the rats exposed to bright **1**_____ showed significant improvement on the spatial **2**_____. Further, when the **3**_____ that had been exposed to dim light were then exposed to **4**_____ light for four weeks their brain capacity and task performance recovered fully.

The study is the first to show that changes in environmental light, in a range experienced by humans, leads to structural changes in the brain. Americans spend about 90 percent of their time indoors. "When we exposed the **5**_____ to dim light, mimicking the cloudy days of Midwestern winters or typical indoor lighting, the animals showed impairments in spatial learning," said Antonio Nunez, co-investigator on the study. "This is similar to when people can't find their way back to their cars in a parking lot after being in a movie theater."

Joel Soler, lead author of the paper published in the journal *Hippocampus*, said sustained exposure to **6**_____ light led to significant reductions in a substance called brain derived neurotrophic factor -- a peptide that helps maintain healthy connections and neurons in the hippocampus -- and in dendritic spines, or the connections that allow neurons to "talk" to one another. "Since there are fewer connections being made, this results in diminished learning and memory performance that is dependent upon the **7**_____," Soler said. "In other words, dim lights are producing dimwits."

Interestingly, light does not directly affect the hippocampus, meaning it acts first on other sites within the **8**_____ after passing through the eyes. The research team is investigating one potential site in the rodents' brains -- a group of neurons in the hypothalamus that produce a peptide called orexin that's known to influence a variety of brain functions. One of their research questions: If orexin is given to the rats that are exposed to dim light, will their brains recover without being re-exposed to **9**_____ light?

The project could have implications for the elderly and people with glaucoma, retinal degeneration or cognitive impairments. People with eye disease don't receive much light. Scientists may be able to directly manipulate this group of **10**_____ in the brain, bypassing the eye, and provide them with the same benefits of bright light exposure. Another possibility is improving the cognitive function in the aging population and those with neurological disorders. The scientists hope they can help such patients recover from their impairments or prevent further decline.

Dogs Are More Expressive When Someone Is Looking (6)

Directions: Fill in a word that best completes each sentence.

Scientists at the University's Dog Cognition Centre are the first to find clear evidence that dogs move their faces in direct response to human attention. Dogs don't respond with more facial expressions upon seeing tasty food, which suggests that **1**_____ produce facial **2**_____ to communicate, not just because they are excited. Brow raising, which makes the eyes look bigger -- so-called puppy dog eyes -- was the dogs' most commonly used expression in this research.

Dog cognition expert Dr Juliane Kaminski led the study, which is published in *Scientific Reports*. She said: "We can now be confident that the production of **3**_____ expressions made by dogs are dependent on the attention state of their audience and are not just a result of dogs being excited." The findings appear to support evidence dogs are sensitive to humans' **4**_____ and that expressions are potentially active attempts to communicate, not simple emotional displays.

The researchers studied 24 dogs of various breeds, aged one to 12. All were family pets. Each dog was tied by a lead a meter away from a person, and the dogs' faces were filmed throughout a range of exchanges, from the **5**_____ being oriented towards the dog, to being distracted, and with the person's body turned away from the dog. The dogs' facial expressions were measured using DogFACS, an anatomically based coding system which gives a reliable and standardized measurement of facial changes linked to underlying muscle movement.

Co-author and facial expression expert Professor Bridget Waller said "DogFACS captures movements from all the different **6**_____ in the canine face, many of which are capable of producing very subtle and brief facial movements." FACS systems were originally developed for humans but have since been modified for use with other animals such as primates and dogs.

Dr Kaminski said: "Domestic dogs have a unique history -- they have lived alongside **7**_____ for 30,000 years and during that time selection pressures seem to have acted on dogs' ability to **8**_____ with us." It is possible dogs' facial expressions have changed as part of the process of becoming domesticated.

It is impossible yet to say whether dogs' behavior in this and other studies is evidence dogs have flexible understanding of another individual's perspective -- that they truly understand another **9**_____ mental state -- or if their behavior is hardwired, or even a learned response to seeing the face or eyes of another individual.

Previous research has shown some apes can also modify their facial expressions depending on their audience, but until now, dogs' abilities to do use facial expression to communicate with **10**_____ hadn't been systematically examined.

Driving Speed Affected When a Driver's Mind "Wanders" (7)

Directions: Fill in a word that best completes each sentence.

Research from North Carolina State University finds that driving speed fluctuates more when a driver's mind wanders from focusing on the act of **1**_____ -- and that the outside environment influences how often a driver's **2**_____ wanders.

"As automatic and semi-automatic technologies take over some driving tasks, drivers are likely to experience increased boredom because they will have less to **3**_____, which makes it important to understand what contributes to 'mind wandering' for drivers, and how that affects driver performance," says Michal Geden, a Ph.D. candidate at NC State and the lead author of a paper on the work. Mind wandering is thinking about anything other than driving.

The more drivers are bored because they have less to concentrate on while they drive, the more their minds **4**_____. This study tells us that mind wandering causes variations in driving speed, which effects safety.

Additionally, if there is less information in the environment that the driver needs to process, their mind will also tend to wander more. For example, driving in a low population desert **5**_____ will cause more mind wandering than driving in a busy, urban area.

For this study, researchers conducted an experiment with 40 drivers using a driving simulator. A driving simulator is a **6**_____ model that can be used to act out real-life **7**_____ conditions. Drivers reported that their minds wandering 50 percent of the time in environments with little activity (like a low population desert). In **8**_____ with a lot of activity, like a busy, **9**_____ area, drivers reported mind wandering only 41 percent of the time. During times that drivers reported mind wandering, researchers saw increased variations in driving speed.

"There is a great deal of research on external distractions, such as talking on your **10**_____ while driving," says Jing Feng, an assistant professor of psychology at NC State and one of the authors of the paper. "However, not much work has been done on allowing one's mind to wander while driving. As technologies take over more driving tasks, we may become more likely to let our minds wander. We should know what that might mean for vehicle safety, and this study is one step in that direction."

NIH Ends All Research on Chimps (8)

Directions: Fill in a word that best completes each sentence.

The U.S. National Institutes of Health (NIH) recently ended all its funding for research using chimpanzees, going further than its previously stated goal to keep 50 chimps for experiments (per a 2013 report). According to a public release from the agency signed by NIH director Francis H. Collins, he has "reassessed the need to maintain **1**_____ for biomedical research and decided that effective immediately, NIH will no longer maintain a colony of 50 chimpanzees for future **2-**_____."

All the 390 remaining chimps the NIH owns will be headed to a federally-funded chimpanzee sanctuary called Chimp Haven located in Keithville, Louisiana. Chimp **3**_____ has been operating for 20 years. The **4**_____ contains over 200-acres of forest where the chimps can roam free, the Shreveport Times previously reported.

NIH's Collins cautioned in his statement that the transition would take time: "Relocation of the chimpanzees to the Federal Sanctuary System will be conducted as space is available and on a timescale that will allow for optimal transition of each individual chimpanzee with careful consideration of their welfare, including their health and social grouping."

Collins also noted the decision impacted only the chimpanzees, not any of the other species (including up to 25 million mice, rats, pigs, dogs) involved in research projects throughout the country. The conclusion to his release reads: "These decisions are specific to chimpanzees. Research with other non-human primates will continue to be valued, supported, and conducted by the **5**_____."

Collins decision regarding continued research on other **6**_____ was criticized by the pro-animal research activism group Speaking of Research and The Humane Society. Intriguingly, his announcement also came after a report by *Science Insider* covering an aggressive and possibly harassing campaign waged by the animal activism group People for the Ethical Treatment of Animals (PETA), which targeted Collins and NIH researcher Stephen Suomi. Suomi's Maryland, laboratory had been doing research on monkeys (not chimpanzees).

Suomi's research involved separating **7**_____ from their mothers and getting them addicted to alcohol. Suomi's research prompted an ethics inquiry from members of Congress. PETA mailed letters to Collins's and Suomi's neighbors in Washington, D.C. saying the experiments were "cruel," warning the **8**_____ it was "similar to having a sexual predator in your neighborhood," and including Collins' and **9**_____ home addresses and phone numbers. Collins didn't specify whether the campaign had anything to do with his **10**_____ regarding the ending of NIH research on chimpanzees.

Even Insects Have Distinct Personalities (9)

Directions: Fill in a word that best completes each sentence.

Every pet owner knows that their dog, cat or guinea pig has its own individual and distinct personality. Now research has backed this up, showing that even insects have character traits. Personality, or the characteristic way a person is most likely to behave in any situation, is generally considered to be something human - hence 'person-ality'. But research is now finding that animals can have very **1**_____ and individual personalities. Scientists believe this is important for species survival.

Dr von Merten, lead researcher on one EU-funded project looked at **2**_____ personality differences between shrews. She examined behavioral traits such as boldness and shyness, how likely a shrew is to explore a new environment, how friendly a shrew might be and how likely a **3**_____ is to pick a fight. To do this, the researchers placed shrews in small boxes with wooden walls.

To investigate boldness, the researchers placed the box in an open empty cage. The door to the box was opened, and the researchers measured how long the animal took to leave the **4**_____. To measure aggressiveness, the researchers measured how often the shrews bit and struggled during handling or stayed calm and conserved energy as they waited to be released.

Another EU-funded project examined personality in bumblebees. Two **5**_____ traits studied were learning speed and preferences for different colors. The researchers conducted their study by creating an artificial garden where all the 'flowers' were a slightly different color. Learning speed was measured by observing which bees quickly determined the position of the 'flower' with the reward.

Even though they had been working separately, the scientists studying the shrews and the **6**_____ found a range of differences between individuals within each species. These characteristics form distinct personalities. "Each individual has its own way of solving problems," Professor Lars Chittka, the scientist who oversaw the bee project, explained. "Individuals that are highly accurate are also careful and slow in making decisions, whereas other individuals that are faster at making **7**_____ but sloppy."

"People always think that if it's an animal, then there should be only one beneficial way to act, and they should not vary from this," said Dr von Merten. "But for the **8**_____, it's often beneficial to have a variety of different personalities to choose from." For example, in a group, the bold and inquisitive individuals might forage and hunt better than their shyer counterparts. But if predators suddenly moved into an area, then the bold and **9**_____ animals would be the first to be eaten, because of their inquisitiveness, while the more cautious individuals would be more likely to escape and survive.

The more social a species, the more likely there will be variation in personality. In a beehive, having a **10**_____ of different personalities is good because it means labor can be divided effectively and the hive will run better. With shrews it's better to have a social hierarchy to avoid constant fighting.

Freezing Hunger-Signaling Nerve May Help Ignite Weight Loss (10)

Directions: Fill in a word that best completes each sentence.

Freezing the nerve that carries hunger signals to the brain may help patients with mild-to-moderate obesity lose weight, according to a study presented at the Society of Interventional Radiology's 2018 Annual Scientific Meeting. The treatment was determined safe and feasible in the initial pilot phase.

"We developed this treatment for patients with mild-to-**1**_____ obesity to reduce the attrition [stopping or giving up] that is common with weight-loss efforts," said David Prologo, an interventional radiologist from Emory University School of Medicine, and lead author of the study. "We are trying to help **2**_____ succeed with their own attempts to lose **3**_____."

During the procedure, an interventional **4**_____ inserts a needle through the patient's back and, guided by live images from a CT scan, uses argon gas to freeze the nerve, known as the posterior vagal trunk. This nerve, located at the base of the esophagus, is one of several mechanisms that tells the brain that the stomach is empty.

In the study, 10 subjects with a Body Mass Index (BMI) between 30 and 37 underwent the procedure and were followed for 90 days. All subjects reported decreased appetite. The overall average weight loss was 3.6 percent of initial body weight, which represented an average decline of nearly 14 percent of the excess BMI. No procedure-related complications were reported, and there were no adverse events during the follow up.

"Medical literature shows the vast majority of **5**_____-loss programs fail, especially when people attempt to reduce their food intake," said Prologo. "When our stomachs are **6**_____, the body senses this and switches to food-seeking survival mode. We're not trying to eliminate this biological response, only reduce the strength of this signal to the **7**_____ to provide a new, sustainable solution to the difficult problem of treating mild **8**_____."

Following the success of this preliminary safety and feasibility study, more patients will be recruited. The larger clinical trial of the **9**_____ will test its efficacy (effectiveness) and determine how long the effect of the procedure lasts. In presenting the study, the authors note several limitations, including the small sample **10**_____ and the interim (short time span) nature of the results.

The study was funded by HealthTronics, a medical technology company that manufactures the ablation probes used for the treatment.

Bonobos Prefer Hinderers over Helpers (11)

Directions: Fill in a word that best completes each sentence.

Humans generally prefer people who are nice to others. A 2007 study showed that even humans as early as the infant stage prefer helpers. "It was striking and unexpected and suggested that these sorts of motivations may be really central to humans' unusually cooperative nature," Duke University scientist Christopher Krupenye said. He and fellow researcher Brian Hare wanted to find out whether the motivation to prefer **1**_____ might be unique to humans.

In the animal kingdom, chimpanzees and bonobos are humans' closest relatives. Bonobos have been shown to be less aggressive than **2**_____ so Krupenye and Hare expected them to prefer helpers like humans. But the researchers were wrong. It turns out that **3**_____ are more attracted to jerks!

The research team studied adult bonobos at Lola ya Bonobo Sanctuary in the Democratic Republic of Congo. In one series of trials, the researchers showed 24 bonobos videos of a Pac-Man-like shape as it tries to climb a hill. Then another cartoon shape enters the scene. Sometimes it's a helpful character who gives the Pac-**4**_____ a push up, and other times it's an unhelpful one who shoves him down. Afterwards, the scientists offered the bonobos two pieces of apple, one placed under a cutout of the **5**_____ character and another under the unhelpful one and gauged their preference by watching to see which they reached for first.

In another experiment, the bonobos watched a skit in which an actor drops a stuffed animal out of reach. Then another person tries to return the toy to its owner, but before they can a third person snatches it away. Afterwards, the bonobos choose whether to accept a piece of **6**_____ from the do-gooder or the thief.

In each experiment, the bonobos were able to distinguish between helpful and **7**_____ individuals just like humans can. But unlike **8**_____, most bonobos tended to choose the jerks. "There may be a good reason for these puzzling results. It could be that bonobos interpret rudeness as a sign of social status and are simply trying to keep dominant individuals on their side. In other words, it pays to have powerful allies [friends]. For bonobos, making friends with **9**_____ individuals could mean better access to food, mates or other perks, or less chance of being bullied themselves," Krupenye said.

Bonobos are highly socially tolerant in food settings and help and cooperate with food in ways that we don't see in chimpanzees. However, dominance still plays an important role in their lives. The research findings suggest that humans' preference for helpers evolved after our species diverged from other apes. "This **10**_____ may have provided the foundation for the development of complex features of human cooperation," the study authors said. "We continue to explore preferences and social evaluation in bonobos, trying to understand what types of social information they track and what motivates their preferences," he added. "We also plan to conduct similar studies in chimpanzees."

No Evidence to Support Link Between Violent Video Games and Behavior (12)

Directions: Fill in a word that best completes each sentence.

In a series of experiments, done at the University of York, a research team demonstrated that video game concepts don't 'prime' players to behave in certain ways and that increasing the realism of violent **1**_____ games doesn't necessarily increase aggression in players. Previous studies on this topic used smaller sample sizes. These researchers studied 3000 adult participants and compared different types of gaming realism.

The dominant model of learning in video games is built on the idea that exposing players to concepts, such as violence, makes those concepts easier to use in the 'real world'. This is known as 'priming' and is thought to lead to changes in behavior. Previous experiments on this effect have shown mixed results.

In one part of the York study, participants played a **2**_____ where they had to either be a car avoiding collisions with trucks or a mouse avoiding being caught by a cat. Following the game, the players were quickly shown various images, such as a bus or a dog, and asked to label them as either a vehicle or an animal. Researcher Dr David Zendle said: "If players are 'primed' through immersing themselves in the concepts of the game, they should be able to categorize the objects associated with this game more quickly in the real world once the game had concluded. Across the two games we didn't find this to be the case. Participants who played a car-themed game were no quicker at categorizing vehicle images, and indeed in some cases their reaction time was significantly slower."

In a separate part of the York **3**_____, the team investigated whether realism influenced the aggression of game players. Research in the past has suggested that the greater the **4**_____ of the game, the more primed players are by violent concepts, leading to antisocial effects in the real **5**_____. One way violent games can be realistic is the way characters behave. Dr Zendle explained that the experiment looked at the use of 'ragdoll physics' in game design. Characters were modelled after how an actual human skeleton would fall if it was injured.

The experiment compared player reactions to two combat games, one that used 'ragdoll **6**_____' to create realistic character behavior and one that did not, in an animated world that looked real. Following the game, the players were asked to complete word puzzles called 'word fragment completion tasks', where researchers expected more violent word associations would be chosen for those who played the **7**_____ that employed more realistic behaviors. However, Dr Zendle and his team found that the priming of **8**_____ concepts, as measured by how many violent concepts appeared in the word fragment completion task, was not detectable. There was no difference in priming between the game that used 'ragdoll physics' and the game that didn't.

Dr **9**_____ said that further study is now needed on **10**_____ game realism with extreme content, such as torture. The effects of video games on players who are children should also be examined.

Music Taste Linked to Brain Type (13)

Directions: Fill in a word that best completes each sentence.

Scientists at Cambridge University have developed a test that matches a person's personality traits including empathy and systematic thinking with the music they like. David Greenberg and his team wanted to look beyond personality **1**_____ that are used for identifying musical preferences and investigate whether the way a person thinks is a better predictor of their taste in **2**_____.

Candidates first took a questionnaire that looked to identify their brain type - either as an empathizer - someone that feels very strong emotions and can relate to other people's emotions - or a systemizer - someone who thinks very logically and spots patterns in the world. The score from each test was then compared to the baseline for an average person of their gender and their **3**_____ type assigned. A high score on the systemizer quotient, coupled with a low score on the empathizer scale, would represent a "type S" brain; the opposite would be type E. The third brain type, B (balanced) occurs when there is no clear preference for either a systemizer or an empathizer.

Fifty musical excerpts, each around 30 seconds in length, were then played to the candidates and they indicated their preference for each piece of music on a scale from 1 to 9. This was one of the novel methods used in the study, rather than having candidates rank a list of artists or genres. "The problem with genres is that they're so vast. If you take the rock genre in general you have Metallica to Jeff Buckley, so we thought a more accurate way of doing it would just be to administer pieces of **4**_____," explains Greenberg.

The results from the study were very consistent, with a clear difference between the musical **5**_____ of an **6**_____ and those of a **7**_____. "In general empathizers like more mellow music, from genres such as R&B, adult contemporary and soft rock," Greenberg says. "Systemizers like more intense music from the punk and heavy metal genres."

They also looked at **8**_____ tastes in greater detail, identifying the specific features of the music that the participants preferred. Interestingly, empathizers, they found, like "low energy qualities" in their music: romance, emotional depth, poetic and negative emotions, such as depression. Systemizers, on the other hand, prefer much more intense and strong music that's thrilling, fast and contains positive emotions, such as joy.

When comparing the predictions of music **9**_____ from their study, with those from other personality-based studies, the team found that a person's brain **10**_____, or the way that they think, is actually a better predictor of music tastes, in particular for empathizers. This could mean improved musical recommendations for users of platforms such as YouTube or iTunes. Alongside this, Greenberg believes that this research could also help to improve musical therapy techniques. "We could use these results as a way of teaching children emotional recognition through music".

Our Reactions to Odor Reveal Our Political Attitudes, Survey Suggests (14)

Directions: Fill in a word that best completes each sentence.

A recent survey showed that people who are easily disgusted by odors are also drawn to authoritarian political leaders. These people are more sensitive to smells like sweat or urine. These sensitivities might come from a deep-seated instinct to avoid infectious diseases.

"There was a solid connection between how strongly someone was **1**_____ by smells and their desire to have a dictator-like leader who can suppress radical protest movements and ensure that different groups "stay in their places." That type of society reduces contact among different groups and, at least in theory, decreases the chance of becoming ill," says Jonas Olofsson, who researches scent and psychology at Stockholm University and is one of the authors of the study.

Disgust is an emotion that helps us survive. When people are disgusted, they wrinkle their noses and squint their eyes, decreasing their sensory perception of the world. At its core, disgust is a protection against things that are dangerous and infectious -- things that we want to avoid. The researchers had a hypothesis that there would be a connection between feelings of disgust and how a person would want society to be organized. They thought that **2**_____ with a strong instinct to distance themselves from unpleasant **3**_____ would also prefer a society where different **4**_____ are kept separate.

A second author of the study, Marco Tullio Liuzza from Magna Graecia University of Catanzaro, Italy, explained that people have an emotional reaction to potential pathogens, which can lead to disease. One possible indicator of **5**_____ is body odors. Negative emotional reactions can lead to misinformed attitudes, which can further lead to aggression towards groups perceived as different and a desire to segregate them.

This study involved a large survey given online. Participants rated their disgust for body odors on a specially developed scale. The **6**_____ asked participants to rate not only their feelings about their own body **7**_____, but the body odors of others, as well as their own political views. In the US, questions about how they planned to vote in the presidential race in 2016 were added. "It showed that people who were more disgusted by smells were also more likely to vote for Donald Trump than those who were less sensitive." says Jonas Olofsson. Since **8**_____ Trump has spoken about how different people disgust him and that immigrants spread disease, it fit the researcher's hypothesis that Trump's supporters would also be more easily disgusted.

The results of the study could be interpreted to suggest that authoritarian **9**_____ views are innate and difficult to change. However, Jonas Olofsson believes that they can be **10**_____ even if they are deep seated. "The research has shown that the beliefs can change. If contact is created between groups, authoritarians can change. It's not carved in stone. Quite the opposite, beliefs can be updated when we learn new things."

Poor Grades Tied to Class Times That Don't Match Our Biological Clocks (15)

Directions: Fill in a word that best completes each sentence.

Researchers from UC Berkeley and Northeastern Illinois University have shown that students whose circadian rhythms were out of sync with their class schedules (for example, night owls taking early morning courses) received lower grades due to "social jet lag," a condition in which peak alertness times are at odds with work, school or other demands. In addition to learning deficits, social **1**_____ lag has been tied to obesity and excessive alcohol and tobacco use.

Their findings, published in the journal *Scientific Reports*, indicate the researchers tracked the personal daily online activity profiles of nearly 15,000 college students as they logged into campus servers. After sorting the students into "night owls," "daytime finches" and "morning larks" -- based on their activities on days they were not in class -- researchers compared their class times to their academic outcomes.

Researchers Benjamin Smarr and Aaron Schirmer followed student online **2**_____ over two years. They separated the owls, larks, and finches by tracking students' activity levels on days that they did not attend a class. They also looked at how larks, finches and **3**_____ had scheduled their classes during four semesters from 2014 to 2016 and found that about 40 percent were mostly biologically in sync with their class times. As a result, they performed better in class and enjoyed higher GPAs. However, 50 percent of the students were taking classes before they were fully alert, and another 10 percent had already peaked by the time their classes started. "We found that the majority of students were being jet-lagged by their class times, which correlated very strongly with decreased **4**_____ performance," said Smarr, of UC Berkeley.

On a positive note: "Our research indicates that if a student can structure a consistent schedule in which class days resemble non-class days, they are more likely to achieve **5**_____ success," said Schirmer, of Northeastern Illinois University.

While students of all categories suffered from class-induced jet lag, the study found that night owls were especially vulnerable, many appearing so chronically jet-**6**_____ that they were unable to perform optimally at any time of day. But it's not as simple as **7**_____ just staying up too late, Smarr said. "Because owls are later, and classes tend to be earlier, this mismatch hits owls the hardest, but we see **8**_____ and **9**_____ taking later classes and also suffering from the mismatch," said Smarr. "Different people really do have biologically diverse timing, so there isn't a one-time-fits-all solution for education."

The results suggest that "rather than admonish late students to go to bed earlier, in conflict with their biological rhythms, we should work to individualize education so that learning and classes are structured to take advantage of knowing what time of day a given student will be most capable of **10**_____," Smarr said.

This Is Why Science Loves Twins (16)

Directions: Fill in a word that best completes each sentence.

The FBI is interested in us. No, we're not "persons of interest." We are interesting persons. My brother and I are identical twins. And the FBI has been supporting West Virginia University's twin studies here for years. For the last 21 years, this small town in Ohio has lived up to its name (Twinsburg) during the annual Twin Days Festival. This year, we are here with 1,917 other sets of multiples.

The FBI uses **1**_____ to help with facial recognition computer software. One of the greatest tests of a **2**_____ recognition program is telling **3**_____ twins apart. These programs can be made better if they can differentiate between very similar individuals. Facial **4**_____ programs typically help with security in government buildings.

These sorts of measurements of human characteristics aren't just of interest to the U.S government, they are also of interest to Apple. The iPhone X from Apple has a "Face ID" program that makes a 3-D scan of your **5**_____ that you can use to unlock your phone or pay for things. This technology is only going to get more common in the **6**_____.

Face scanning isn't the only way twins are used. Recordings of twins reading the same section in a book can train computers to tell them apart by voiceprint. Science is also interested in looking at the DNA of **7**_____ using their saliva. Twins get tested to see how well they can sense fat, bitter, and sweet tastes. Science is even interested in how twins use social media and what their online news habits are. Hundreds of identical and fraternal twins are sampled, questioned, and scanned to improve scientific understandings.

Although identical twins share 100 percent of their genes, they are exposed to different environments during their **8**_____. These different environments act on the twins in very different ways. If you watch and test how two identical twins grow, age, and ultimately die, you have the best natural experiment for separating the contributions of our **9**_____ and our **10**_____. If researchers can understand "nature vs. nurture" in twins, it will be easier to figure out for all the non-twins out there. And that, said Chance York, an assistant professor at Kent State University, is why researchers make the yearly pilgrimage to Twinsburg.

When we asked him about what science would do if there were no twins to experiment on, he said: "the non-scientific, non-tactful answer is that we'd be screwed."

New Study Finds That More Screen Time Coincides with Less Happiness in Youths (17)

Directions: Fill in a word that best completes each sentence.

Happiness is not a warm phone, according to a new study exploring the link between adolescent life satisfaction and screen time. Teens whose eyes are habitually glued to their smartphones are markedly unhappier, said study lead author and San Diego State University professor of psychology Jean M. Twenge.

To investigate this link, Twenge, along with colleagues Gabrielle Martin at SDSU and W. Keith Campbell at the University of Georgia, crunched data from the Monitoring the Future (MtF) longitudinal study, a nationally representative survey of more than a million U.S. 8th-, 10th-, and 12th-graders. The survey asked students questions about how often they spent time on their phones, tablets and computers, as well as questions about their face to face social interactions and their overall happiness.

On average, they found that teens who spent more time in front of screen devices -- playing computer games, using social media, texting and video chatting -- were less **1**_____ than those who invested more time in non-screen activities like sports, reading newspapers and magazines, and in-person social interaction.

Twenge believes this screen time is driving unhappiness rather than the other way around. "Although this study can't show causation, several other studies have shown that more social media use leads to unhappiness, but unhappiness does not lead to more **2**_____ media use," said Twenge.

Total screen abstinence doesn't lead to happiness either, Twenge found. The happiest teens used digital **3**_____ a little less than an hour per day. But after a daily hour of screen **4**_____, unhappiness rises steadily along with increasing **5**_____ time, the researchers report in the journal *Emotion*.

"The key to **6**_____ media use and happiness is limited use," Twenge said. "Aim to spend no more than two **7**_____ a day on digital media and try to increase the amount of time you spend seeing friends face-to-face and exercising -- two activities reliably linked to greater **8**_____."

Looking at historical trends from the same age groups since the 1990s, the researchers found that the proliferation of screen devices over time coincided with a general drop-off in reported happiness in U.S. **9**_____. Specifically, young people's life satisfaction, self-esteem and happiness plummeted after 2012. That's the year that the percentage of Americans who owned a smartphone rose above 50 percent, Twenge noted.

"By far the largest change in teens' lives between 2012 and 2016 was the increase in the amount of time they spent on digital media, and the subsequent decline in in-person social activities and sleep," she said. "The advent of the **10**_____ is the most plausible explanation for the sudden decrease in teens' psychological well-being."

Sorry, Grumpy Cat: Study Finds Dogs are Brainier Than Cats (18)

Directions: Fill in a word that best completes each sentence.

New research offers a twist on the old argument about which is smarter, cats or dogs. The study focuses on how many neurons (grey cells) are in the brains of carnivores. Neurons are associated with thinking, planning and complex behavior -- all considered indicators of intelligence.

Associate Professor of Psychology and Biological Sciences Suzana Herculano-Houzel and her team developed a method for measuring the number of neurons in **1**_____. The results of the study are described in a paper titled "Dogs have the most **2**_____, though not the largest brain: Trade-off between body mass and number of neurons in the cerebral cortex of large carnivoran species" in *Frontiers in Neuroanatomy*.

The team found that dogs have about 530 million neurons while cats have about 250 million. (That compares to 16 billion in the human brain.) Herculano-Houzel explained that the number of neurons an animal has, especially in the cerebral cortex, determines their ability to predict what is about to happen in their environment based on experience.

The brains of one or two specimens from eight different carnivore species were analyzed. The researchers expected that the brains of **3**_____ would have more neurons than the herbivores because hunting requires more thinking than grazing. The data did not support that **4**_____, however. The ratio of neurons to brain size in small- and medium-sized carnivores was about the same as that of **5**_____, suggesting that there is just as much evolutionary pressure on the herbivores to develop the brain power to escape from predators as there is on **6**_____ to catch them. In fact, the largest carnivores have the lowest neuron-to-brain-size ratio. They found that the brain of a golden retriever has more neurons than a lion or brown bear, even though these bigger **7**_____ have much larger brains.

Meat eating provides a lot of energy but there is a delicate balance between the size of the brain and the size of the body, explains Herculano-Houzel. Large predators burn much **8**_____ hunting and it may be a long time between successful kills. That explains why large meat-eating carnivores like lions spend most of their time resting. The brain needs energy continuously; the more **9**_____, the more energy necessary. Consequently, the quantity of meat that large hunters can kill and consume, and the intermittent nature of feeding appears to limit their brain development.

The researchers also found that domestic animals like ferrets, cats, and dogs don't have smaller brains than wild animals like the mongoose, raccoon, hyena, lion and brown bear. The analysis also discovered that the raccoon had a great many neurons in its small brain. Studying the **10**_____ of different species teaches us that there are patterns, but also diversity in the way that nature has found to put animal brains together.

Study of Life

Are Viruses Alive? New Evidence Says Yes (1)

Directions: Fill in a word that best completes each sentence.

Influenza, SARS, Ebola, HIV, and the common cold are all viruses. A virus has genetic material (DNA or RNA) that is encapsulated (surrounded) in a protein coat. Virologists (scientists who study viruses) have struggled with the question of whether viruses are living or not.

Many scientists have long argued that viruses are nonliving. Based on everything known about characteristics of life, viruses don't qualify. There are many life processes, such using energy and nutrition, that viruses do not do. Viruses seem to carry out only one life process: reproduction. Though, even then, individual viruses must invade a cell and hijack its genetic tools in order complete **1**_____.

These current understandings are incomplete, however. Recently, virologists have discovered that some viruses, which have more genetic information than typical viruses, have genes for proteins that can perform translation, allowing them to produce their own offspring. Additionally, another characteristic of **2**_____, evolution, is also performed by **3**_____. Researchers Carl Woese, Gustavo Caetano-Anolles, and Arshan Nasir published their research in *Science Advances*.

To do their research the virologists couldn't focus on viral DNA or RNA because these molecules frequently mutate. Instead they looked at protein folds, three dimensional structures in proteins. These are the puzzle-like shapes of proteins that allow them to perform basic molecular functions. Some folds are shared by all organisms, while others are unique to individual branches of the Tree of Life. The specific shapes are coded by genes, and do not change much over time, unlike DNA or **4**_____ sequences. Protein folds provide a good marker to look back in history.

Research analysis of viruses and organisms from every branch of the **5**_____ of Life showed similarities of 442 protein folds between cells and viruses. Viruses possessed 66 unique folds, which indicates that they may have branched away from other organisms at some point. Viruses must have diversified from ancient cells by a process called reductive evolution, where organisms simplify instead of becoming more complex. At some point, the genetic **6**_____ of these ancient viral cells was reduced to the point where they lost their cellular nature and became modern viruses. Viruses restore their 'cellular' existence today when they enter and take control of any **7**_____. The cells that these viruses lived within were the last common ancestor, which probably existed 2.45 billion years ago.

Greater understanding of reductive **8**_____ has revealed numerous examples of parasitic organisms like bacteria and fungi that rely on hosts to complete their life cycles. The **9**_____ hope that their findings, added to existing evidence, will prompt the scientific community to include viruses in the picture of cellular evolution, which includes a Tree of Life with **10**_____ for both viruses, as well as cells.

Understanding Bacterial Wargames Inside Our Body (2)

Directions: Fill in a word that best completes each sentence.

Much like animals and, to a degree humans, bacteria enjoy a good fight. They stab, shove and poison each other in pursuit of the best territory. While this much is clear, little is known about the tactics and strategy that **1**_____ use during their miniature wargames.

In a study published in *Current Biology*, researchers at the University of Oxford have revealed that bacteria approach conflict in the same way as an army. The authors of the study examined pairs of *Escherichia coli* strains as they fought against each other. Each strain used a different toxin to try to overcome its competitor. A bacterial strain is immune to its own toxins, but those **2**_____ can kill other bacterial **3**_____. This type of competitive interaction plays a key role in how individual bacteria establish themselves in a community, such as the human gut.

By engineering the strains to have a fluorescent-green color, the research study's authors were able to clearly follow their combat in real time. The findings revealed that not all strains of bacteria fight the same way. Each approaches conflict with a different level of attack, some being hyper-aggressive and others much more passive. In addition to these basic differences in aggression, the research also shows that some strains can not only detect an attack from an incoming toxin, but they can also respond quickly to warn the rest of the colony. Cells on the edge of the **4**_____ will detect the incoming **5**_____ and share this information with the cells behind the battlefront, allowing them to respond with a coordinated, collective retaliation.

Professor Kevin Foster, senior author on the work and Professor of Evolutionary Biology in the Department of Zoology at the University of Oxford, said, "Our research shows that what appear to be simple organisms can function in a very sophisticated manner. Their behavior is more complex than we have previously given them credit for. Much like social insects, such as honey bees and wasps and social animals like birds and mammals who use alarm calls, when under predation, [bacteria] are capable of generating a coordinated attack."

The human body plays host to vast numbers of bacteria, particularly our gut. There is a bacterial war going on inside us! Understanding bacterial competition can help us determine how bacteria spread. Professor Foster explains that understanding that bacteria release toxins in a sophisticated manner and out-compete others is very important for understanding the spread of **6**_____.

The research team is in the process of building on this work to understand how bacteria can use **7**_____ to provoke and misdirect aggression toward their opponents. Dr Despoina Mavridou, one of the lead authors on the study, explained that warfare based on provocation can be beneficial. **8**_____ is most likely taking place in the human **9**_____, where bacteria may **10**_____ multiple opponents to attack and wipe out each other.

Bacteria on Kitchen Towels (3)

Directions: Fill in a word that best completes each sentence.

Scientists interested in the presence of bacteria found in kitchens undertook a project of sampling common kitchen dish towels. The lead study author was Dr. Susheela Biranjia-Hurdoyal, a senior lecturer in the Department of Health Sciences at the University of Mauritius. The **1**_____ was presented at the American Society for Microbiology in Atlanta. The researchers involved in the study cultured bacteria which had been gathered from 100 unwashed kitchen towels. They found that 49 percent of the towels, which each had been used for one month, tested positive for **2**_____.

The number of bacteria present was higher for towels used by large families or families with children, compared with towels used by smaller **3**_____ or families without **4**_____. Towels used for multiple purposes, such as wiping utensils, drying hands, and wiping surfaces grew more bacteria than towels used for a single purpose. The researchers also found that damp towels grew more bacteria than dry towels. They recommended against multipurpose usage of kitchen towels and said that larger families "should be especially vigilant [when it comes] to hygiene in the kitchen."

Of the towel samples that tested **5**_____ for bacteria, 73 percent grew types of bacteria found in human intestines, including *E. coli* and Enterococcus species. About 14 percent grew *Staphylococcus aureus*, or staph, a bacterium that is sometimes found on people's skin. Although staph bacteria usually don't cause **6**_____ in healthy people, when the bacterium gets into food, it can produce toxins that can cause food poisoning.

Benjamin Chapman, an associate professor and food safety specialist at North Carolina State University, said the study gives us a look at what bacteria are in the environment around us. But "it doesn't surprise me at all that something that's in a kitchen environment has bacteria on it. We really do live in a world that's dominated by microorganisms," said Chapman.

These findings may sound gross, but the bacteria found on the towels in this study aren't particularly concerning when it comes to foodborne illnesses. Chapmans says, "what's listed here doesn't initially raise concerns with me." The study didn't find any of the common culprits of **7**_____ illness, such as Salmonella, Campylobacter or pathogenic types of *E. coli*, such as *E. coli* O157:H7. (*E.* **8**_____ O157:H7 can infect people who consume contaminated food, raw food or raw (unpasteurized) milk. Infection with this type of pathogenic bacteria may lead to hemorrhagic diarrhea, and to kidney failure.)

Chapman says that, in theory, **9**_____ towels could aid in the spread of foodborne **10**_____. This could happen if, for instance, someone used a kitchen towel to wipe up meat juices from the counter and another person unknowingly used the towel to dry their hands. To avoid this, Chapman recommends frequently washing and drying kitchen towels to prevent bacterial growth.

Rabbit, Dog, Human: How One Bacterial Infection Spread (4)

Directions: Fill in a word that best completes each sentence.

A woman in Arizona died from an infection called rabbit fever, despite never coming into contact with any **1**_____, according to a recent report of the woman's case.

The 73-year-old woman first got **2**_____ on June 6, 2016, and died five days later from severe breathing problems, according to a report published today by the Centers for Disease Control and Prevention. It wasn't until June 17 of that year, however, when the results of a blood test came back, that doctors learned the woman had rabbit fever, which is also called tularemia.

Rabbit fever is a bacterial infection caused by the bacterium *Francisella tularensis*, according to the report. **3**_____ typically start three to five days after exposure to the **4**_____ and can include fever, skin lesions, difficulty breathing and diarrhea. Though the infection can be deadly, most infections can be treated with **5**_____, according to the CDC.

People can get rabbit fever through insect bites, coming into contact with an infected animal or inhaling the bacteria.

Though the woman lived in a semirural area, she told doctors that she didn't participate in outdoor activities, according to the report. In addition, the woman didn't have any insect bites, and hadn't been exposed to any animal carcasses or untreated water, the report said.

Her dog, however, had been found that May with a dead rabbit in its mouth, and was later noted to be lethargic and eating less. After the woman died, doctors tested the **6**_____ and found signs of the infection in its blood. In addition, investigators found a few infected **7**_____ around the woman's property.

Because the woman had respiratory symptoms, the researchers think she inhaled the **8**_____, potentially from her dog, the report said. It's possible that the dog had the bacteria in its mouth after catching the dead **9**_____, or there were bacteria on its fur, the authors said.

About 125 rabbit **10**_____ cases are reported in the U.S. each year, the report said.

Cellular Messengers Communicate with Bacteria in the Mouth (5)

Directions: Fill in a word that best completes each sentence.

Dr. David Wong, of the UCLA School of Dentistry, and Dr. Wenyuan Shi, of the Forsyth Institute, an oral health research institute in Massachusetts, wondered if human RNA can communicate with harmful bacteria in the mouth. Their research question and the study that followed it has been published in the *Journal for Dental Research.*

RNA acts as a chemical messenger that transports DNA's instructions to other parts of the cell. There are different kinds of RNA. One non-coding type of **1**_____ is sRNA, which regulates our genes. A class of sRNA, called tsRNA, is found in human body fluids, including blood, tears and saliva.

The researchers analyzed saliva and found many tsRNA sequences that matched transfer RNA (tRNA) sequences of Gram-negative oral bacteria. Gram-**2**_____ oral **3**_____ have a toxic outer layer that can cause periodontal disease. Since there is a match, the salivary tsRNA could affect bacterial tRNA protein synthesis and cause more bacterial growth, leading to periodontal **4**_____.

The **5**_____-negative (antibiotic-resistant) bacterium used to test the researcher's hypothesis was *Fusobacterium nucleatum* (*F. nucleatum*), the bacteria responsible for periodontitis. The research team showed that salivary host cells respond to the presence of *F.* **6**_____ by releasing a sequence of tsRNA which matched the tRNA of **7**_____ *nucleatum*. Furthermore, although the salivary tsRNAs can communicate with the tRNA of gram-negative *F. nucleatum*, they have no effect on the gram-positive oral bacteria *Streptococcus mitis*.

"This study establishes that there is a clear channel of communication between RNA messengers and bacteria in our mouth," said Wong. Another significant study finding was the majority of tRNA bacteria sequences that show high sequence similarity with **8**_____ tsRNAs came from **9**_____-resistant Gram-negative bacteria. This observation could lead to a better understanding of the mechanisms behind the growth of oral bacteria, resistance to antibiotics, and in-turn oral diseases, Wong said.

"Our findings could lead to new therapies to treat diseases caused by harmful bacteria," said Shi. For example, one of the hallmarks of periodontitis is a shift from mostly Gram-**10**_____ bacteria to mostly Gram-negative bacteria. With a better understanding of how Gram-negative bacteria grow, perhaps there is potential to reverse the growth or even kill Gram-negative bacteria.

Fecal Transplantation (6)

Directions: Fill in a word that best completes each sentence.

Fecal transplants are increasingly being used as the treatment for infections in the human gut. They have had success treating the *Clostridium difficile* colitis, an infectious diarrhea that often follows antibiotic treatment. There has also been an increase in animal experiments involving fecal material. In one study, for example, researchers found that fecal **1**_____ from donor mice were able to make recipient mice either lean or fat.

"This research is just getting started. It is driven by the new paradigm of the microbiome which recognizes that every plant and animal species harbors a collection of microbes that have significant and previously unrecognized effects on their host health, evolution and behavior," said Seth Bordenstein, associate professor at Vanderbilt University.

In an article titled "**2**_____ Transplants: What is Being Transferred" published in the journal *PLOS Biology*, Bordenstein reviews the scientific literature, which shows that fecal transplants to treat **3**_____ *difficile* infections have a 95 percent cure rate. "Right now, fecal transplants are used as the treatment of last resort, but their effectiveness raises an important question: When will doctors start prescribing them, or some derivative, first?" Bordenstein asked.

Fecal transplantation has a long history. In the 4th century, a Chinese medical doctor named Ge Hong used fecal **4**_____. In 16th century Chinese Medicine it had the nickname "yellow soup." Western medicine recently began to show interest. The trend is expected to rise.

Most published research on fecal transplantation has focused on the role of the bacteria in donor's stool. **5**_____ are the most abundant active agent in the material. However, they are not the only functional ingredient in feces, Bordenstein cautioned. "Feces is a complex material that contains a variety of biological and chemical entities that may be causing or assisting the effects of these transplants." Healthy human stool contains on average 100 billion bacteria per gram. But it also contains 100 million viruses and archaea per gram. (Archaea are single-celled organisms that were classed as bacteria until the 1970's). In addition, there are about 10 million colonocytes (human epithelial cells that help protect the colon) and a million yeasts and other single-celled fungi per **6**_____.

According to **7**_____, focus on the bacterial component appears to make sense in some cases, but in other **8**_____, such as the possible treatment of multiple sclerosis, it is possible that the effects of fecal transplants may be influenced by, or caused by, their non-**9**_____ components. As a result, he calls for more research to determine the effects and interactions of each of these **10**_____. "When scientists identify specific cocktails that produce the positive outcomes, they can synthesize them and put them in a pill. That will help to reduce the 'icky factor' that slows public acceptance of this treatment," said Bordenstein.

Good Germs Can Be Found in Poop (7)

Directions: Fill in a word that best completes each sentence.

Ari Grinspan, a doctor at Mount Sinai Medical Center, regularly uses poop as medicine. The procedure, called a fecal transplant, isn't for everyone. But "in the right patient, in the right setting," he says, **1**_____ can improve a person's health. He injects fecal material directly into a patient's intestines, or bowels. Some transplants use a tube placed down a person's throat. Some patients swallow the **2**_____ material in a special drink or as pills. Thomas Borody, of the Centre for Digestive Diseases, jokingly calls the pills "crapsules." Swallowing **3**_____ containing fecal **4**_____ can be easier for patients than getting fecal material injected into their bowels.

Human feces contain trillions of bacteria. These tiny germs live inside us, forming communities called microbiomes. Each **5**_____ is a miniature ecosystem full of different species. These communities exist on human guts, skin, noses and elsewhere. They help keep the body healthy.

Sometimes, though, dangerous microbes invade or take over. When they do, they cause infection, or make chemicals that harm other systems in the body. In both cases, the human host suffers. In many diseases, the gut microbiome is out of homeostasis. Getting it back to normal might treat or cure the disease. One way to do that is to put a sample of good bacteria from a healthy person's feces into the sick person's **6**_____.

The healing power of poop is not a new idea. It has been used as a part of Chinese medicine for centuries. However, Western medicine only uses it as treatment for one condition: *Clostridium difficile* colitis. *Clostridium* **7**_____ colitis causes diarrhea and extreme pain in the gut. A fecal **8**_____ is a safe and effective cure for this disease.

As researchers more thoroughly learn about microorganisms, the medical community has been finding that they need to give good germs more respect. Besides gut **9**_____, other microbes living in the human body are also likely produce powerful antibiotics. This explains why fecal transplants work. A healthy human gut already contains substances that fight off bad bacteria.

Many people fear germs of all kinds. They scrub their hands and household surfaces regularly with germ-killing products. They avoid dirt or sanitize their hands after petting a dog or cat. Researchers are now realizing that there's such a thing as being too clean. The body needs a diverse community of microbes to stay healthy. "If a dog licks you, I would not use a sanitizer," says Anita Kozyrskyj, an epidemiologist (a doctor who studies patterns of disease), "You can wash it off with water." Grinspan agrees that it's a good idea to avoid using antibacterial soaps. These products destroy good bacteria along with the bad. He also says, "I tell my kids that playing in the dirt is okay." So, next time you flush the toilet, pet a dog or blow your nose, think of the good germs growing in and on your **10**_____. Then wash your hands — but stick to regular soap.

How Often Should I Clean My Phone? (8)

Directions: Fill in a word that best completes each sentence.

The world is a giant petri dish. Everything is carpeted with microorganisms. The good news is that most them are either benign or beneficial. That's true even of fecal bacteria. While a sick person's excrement harbors illness-causing germs, a healthy person's poop—though gross—usually isn't dangerous. (In fact, fecal transplants can confer health benefits.) Your smartphone is also enveloped in **1**_____, most of which pose no threat. "It's unusual that the general bacteria healthy people leave on surfaces like our phones will make us sick," says Emily Toth Martin of the University of Michigan.

That said, germs that can sicken you are out there. And because many of us touch our phones while we eat meals, extra caution is warranted when it comes to these devices. Illness-causing **2**_____— including the types that cause food poisoning, common colds and other infections—can only make you **3**_____ if they enter your body, says Philip Tierno, a clinical professor of microbiology and pathology at New York University. They do that by clinging to hands and then sloughing off into the mouth, ears, nostrils or breaks in the skin. Cellphones can also be a "vehicle that can effectively transfer infectious organisms," Tierno says.

Assuming you don't have open wounds on your hands, "it doesn't matter how dirty they become if you **4**_____ them well with soap and water before you eat," Tierno says. But if, after washing, you pick up a germ-ridden phone—the same one you consulted at the grocery store while you were shopping for raw meat—you're putting yourself in jeopardy. Tierno adds, "The more people touch a surface, the greater the risk of contamination and sickness if you touch that **5**_____ too."

How often you should clean your **6**_____ depends on where you've been and how you handle your device. If you never use your phone while eating, you don't have to be as diligent about cleaning your device. But if you tend to use your phone all the **7**_____—including during meals—a daily cleanse with a disinfectant wipe is a good idea. "I **8**_____ mine twice a day, once in the morning and once in the evening," Tierno says.

Regular cleanings may be especially beneficial if you use a rubber phone cover; bacteria tend to more easily cling to that material than to your phone's metal, glass or hard plastic components, he says. Also, the raised edges where your phone and protective case come together can trap bacteria more effectively than a phone's smoother surfaces, so those crannies warrant extra attention when you disinfect.

If you're worried about hurting your device, Apple and other manufacturers recommend turning off your phone before **9**_____ and avoiding getting liquid into the device's charging port. Either use a pre-treated disinfectant wipe to clean it or leave your phone in your pocket and wash your hands before eating. That will prevent bacteria from the phone from getting into your **10**_____.

Mummies' Tummies to Reveal Digestive Evolution (9)

Directions: Fill in a word that best completes each sentence.

Mummified bodies from Egypt and the Canary Islands (off the northwestern coast of Africa) are having their digestive tracts tested and compared to living people to reveal how gut bacteria have changed over the centuries and how they vary according to diet. It is thought that the more diverse the mix of bacteria in your **1**_____, the healthier you are.

Dr Ainara Sistiaga has been examining mummies to better understand the micro-biome of our ancestors. Little is known about how gut microbes changed during the evolution of humans as hunter-gatherers, to farmers, to processed food eaters. It could be beneficial to know what sorts of **2**_____ lived inside human ancestors and what gut bacteria have been lost, says Dr Sistiaga. The **3**_____ that Dr Sistiaga has focused on include those preserved on the Canary Islands and **4**_____ from before the 15th century. When alive, some of these people were pastoralists (animal herders), others were farmers, and some had been desert dwellers.

The human gut carries trillions of individual microbial cells and is essential to human health. Scientists have learned that the **5**_____ gut can either host a diverse and healthy collection of intestinal bugs or a more minimalist, unhealthy collection. A high-fat Western diet can encourage an unhealthy micro-biome, which often is found in those who are overweight. Worryingly, this situation is linked to diabetes, high blood pressure, inflammation and cardiovascular disease.

Conversely, diets high in fruit and vegetables result in the growth of an intestinal bacterium called *Akkermansia muciniphila.* This bacterium makes up 5 % of bugs in a healthy gut. In overweight people, this bacterium can be almost entirely absent. Interestingly, when **6**_____ *muciniphila* is fed to obese mice, it reduces their weight gain, cuts down on their bad cholesterol, and lessens inflammation. Human studies hint at benefits too: obese people with more of this bacterium in their gut at the start of a six-week diet displayed better metabolic and heart health readings at the end.

Akkermansia **7**_____ provides health benefits, while unhealthy microbes cause damage. **8**_____ microbes chip away at the human gut barrier. A high-fat, low-fiber diet changes the gut micro-biome and causes leakage of some bacteria and pro-inflammatory compounds into the blood. The weakened gut **9**_____ allows compounds to leak into the blood that ramp up inflammation and are linked to metabolic syndrome: high blood sugar, abnormal cholesterol and high body fat. This increases your risk of heart disease, stroke and diabetes.

Anything that lessens this unhealthy cycle would be a huge gain for the health of humans. That's why this research focusing on the gut **10**_____ of early people can help people living today.

New Antibiotic Found in Human Nose (10)

Directions: Fill in a word that best completes each sentence.

A new study, reported in *Nature*, has identified a compound that is produced by one species of nose-dwelling bacterium that kills a different type of bacterium. The study is "yet another demonstration that we should look to nature for solutions to the problems nature throws at us," says Andrew Read, an evolutionary biologist, who was not involved with the work. The **1**_____ that was identified may be developed into an antibiotic, which is good because the world is running out of these life-saving drugs. Researchers behind the new finding believe that studying the microbial warfare going on inside our bodies may lead to not just one, but a whole slew of novel drugs. "We've found a new concept of finding antibiotics," Andreas Peschel, a bacteriologist said. "We have preliminary evidence at least in the **2**_____ that there is a rich source of many others, and I'm sure that we will find new drugs there."

Peschel and Bernhard Krismer, both from the University of Tübingen, analyzed nose secretions (snot) and found that the nose is not a very hospitable niche for microbes. "If I were a bacterium I would not go into the nose," Peschel says. "There is nothing there, simply a salty liquid and a tiny amount of nutrients." Those harsh conditions might lead to competition for resources, the researchers reasoned. That's why they tested what effect a collection of other Staphylococcus species had on *S. aureus*. One **3**_____, *S. lugdunensis*, turned out to be very good at preventing other **4**_____ from growing. The researchers found that the bacterium produced an antibiotic compound and succeeded in synthesizing it in the laboratory. The **5**_____ compound, which they named lugdunin, inhibited *S. aureus* from growing in a petri dish. When applied to the skin of mice infected with *S. aureus*, it improved their infection. It was also effective against antibiotic-resistant strains like MRSA (methicillin-resistant *Staphylococcus aureus*), which kills over 10,000 people a year in the US.

The researchers say that **6**_____ is a powerful enemy of *S. aureus*. "This is extremely exciting as it provides evidence that a microbial war is ongoing in our **7**_____," says Jack Gilbert, a microbial ecologist at the University of Chicago. The research shows "that certain organisms can be leveraged to create novel **8**_____ that could add to our arsenal of weapons against drug-resistant [microbes]." But using such new weapons may have unintended consequences, warns Peer Bork, a computational biologist. "Yes, it's a cool finding," he says. But the microbiome is a delicate balance that has evolved over millions of **9**_____, he cautions. "Tinkering around might destroy long-evolved community relations."

Peschel says, "*S. aureus* is really the most important pathogen that colonizes human noses." He reasons that, if **10**_____ could get rid of it, it would be hard to imagine a negative outcome. He suggests that *S. lugdunensis* be used to colonize patients at risk from *S. aureus*, as a probiotic treatment for the nose. The problem is that *S. lugdunensis* is itself associated with a range of infections—so that strategy could be dangerous. "There may be other bacteria where that is an option," Peschel says.

A 508-Million-Year-Old Sea Predator With a 'Jackknife' Head (11)

Directions: Fill in a word that best completes each sentence.

Paleontologists at the University of Toronto (U of T) and the Royal Ontario Museum (ROM) in Toronto have re-examined a tiny, yet exceptionally fierce, ancient sea creature called *Habelia optata* that has puzzled scientists since it was first discovered more than a century ago. The research, conducted by Cédric Aria, a graduate of U of T and Jean-Bernard Caron, senior curator at the ROM and an associate professor U of **1**_____, is published in BMC Evolutionary Biology.

The 2 cm *Habelia* **2**_____ belongs to the group of invertebrate animals called arthropods, which includes spiders, insects, lobsters and crabs. The extinct animal lived during the middle Cambrian period approximately 508 million years ago. Its fossilized remains can be found in the Burgess Shale deposit in British Columbia. **3**_____ *optata* was part of the "Cambrian explosion," a period of rapid evolutionary change when most major animal groups first emerged in the fossil record.

Like all arthropods, *Habelia optata* has a segmented body with external skeleton and jointed limbs. What remained unclear for decades, however, was the main sub-group of **4**_____ to which *Habelia* belonged. Previous studies had mentioned that *Habelia optata* could be included with mandibulates, organisms which have antennae and mandibles, which could grasp, squeeze and crush their food.

The new analysis suggests that *Habelia optata* was, instead, a relative of the ancestor of all chelicerates, an arthropod sub-group with chelicerae in front of the mouth. **5**_____ are used to cut food.

Aria and Caron analyzed 41 specimens. Their research shows that the well-armored body of *Habelia optata*, was divided into head, thorax and post-thorax. The head of *Habelia* had appendages that allowed it to feel, chew and grasp. "This complex apparatus of **6**_____ and jaws made *Habelia* an exceptionally fierce predator for its size," said Aria.

Surprisingly, although *Habelia* has a closer evolutionary relationship with chelicerates, its unusual characteristics led researchers to compare it with **7**_____ because of how its head functioned. *Habelia* had appendages that allowed it to feel, in a similar fashion as mandibulates use antennae, as well as **8**_____ that assist in the processing of food, as did mandibulates. This similar function between two species that do not have a close evolutionary relationship is called "convergence." Thus, the researchers concluded that *Habelia* was close to the point of divergence between **9**_____ and mandibulates.

The researchers conclude from the outstanding head structure, as well as from well-developed walking legs, that *Habelia optata* were active predators of the Cambrian sea floors. This expands current understanding of ecosystems at the time of the Cambrian **10**_____, showing another level of predator-prey relationship and its determining impact on the rise of arthropods as we know them today.

Peptidoglycan: The Bacterial Wonder Wall (12)

Directions: Fill in a word that best completes each sentence.

Peptidoglycan is a polymer of amino acids (peptido-) and sugars (–glycan) that makes up the cell wall of most bacteria. In fact, the major functional division of bacterial species is based on the structure of the peptidoglycan layer. This functional **1**_____ is apparent when a special staining technique is used which makes the peptidoglycan wall visible.

Bacteriologist named Hans Christian Gram developed a staining technique (now called Gram stain) to visualize bacterial samples. In the 1880s, pneumonia was a large concern. There were three causes: unknown (later identified as viral pneumonia) and two types of bacterial pneumonia caused by either *Streptococcus pneumoniae* or *Klebsiella pneumoniae*. Pneumonia caused by *Streptococcus* is more contagious and develops faster than **2**_____ caused by *Klebsiella*, which tends to only affect the immuno-compromised. Gram's stain, which was fast and definitive, allowed for the three different types of pneumonia patients to be grouped separately from each other, reducing the spread of pneumonia, and, therefore, preventing disease.

The Gram's **3**_____ worked because of the peptidoglycan **4**_____. The thickened peptidoglycan layer in Gram positive cells allows them to retain the stain. They were called 'Gram positive'. The other cells cannot prevent the stain from leeching out because their outer layer is composed of lipopolysaccharides, rather than peptidoglycan. They were called 'Gram negative'.

Peptidoglycan in the cell wall is also vitally important for the way antibiotics work. The bacterial cell wall is defensive. In the microbial world, one of the most important forces acting against the cell size is water. A bacterial cell is like a salty bubble existing in a less salty environment. The **5**_____ salty environment "wants to" even out all the salt concentrations. Therefore, water tends to rush into the cell to dilute its saltiness until it matches that of the **6**_____, or until it bursts and kills the **7**_____. This is the process of osmosis.

Peptidoglycan acts as a physical barrier to prevent the cell from taking on too much **8**_____ and killing itself. If the **9**_____ wall is broken, then the bacterium loses its protective layer and becomes vulnerable to osmosis, causing it to pop. That is how penicillin, the first antibiotic, works. It inhibits the repair of the peptidoglycan layer, which makes it susceptible to osmotic lysis (bursting).

Penicillin is most effective against Gram positive cells because the peptidoglycan is outside cell. Gram negative bacteria have peptidoglycan under a lipid membrane, so it is harder for penicillin to reach it.

In fact, penicillin is so good at killing gram **10**_____ bacteria, that those bacteria have had to evolve to survive. They do this in two ways, they either destroy the penicillin itself or they change the target of penicillin to something penicillin can't recognize. Either way, human use of penicillin to exploit the peptidoglycan wall triggered an arms race with the microbial world.

If You have Pimples You May Need Better Bacteria (13)

Directions: Fill in a word that best completes each sentence.

Even though more than 80% of Americans suffer from acne, which can cause pimples, cysts, and red, inflamed skin, at some point in their lives, the condition is not entirely understood. Past studies have pointed to *Propionibacterium acnes*, a bacterium that lives in the skin's follicles and pores, as a potential cause, but current scientific understanding is still unclear.

Molecular biologist Huiying Li of the David Geffen School of Medicine, and colleagues decided to take a closer look at the microbe. Using over-the-counter pore-cleansing strips, they sampled **1**_____ from the noses of 101 people, 49 of whom had acne and 52 of whom had clear **2**_____. Then, they examined the bacterial DNA, looking for patterns or variations in the microbes' genes that would help them identify specific strains of bacteria.

Whether they had **3**_____ or pimply skin, all the study participants had similar amounts of *P. acnes* living in their pores, but not all the strains were the same. The researchers found several different **4**_____ of the microbe, including 66 that had never been identified before. When they sequenced the genomes of each strain and compared them, they discovered that two of the strains, RT4 and RT5, were found predominantly in people with acne, and that one strain, RT6, was found almost exclusively in **5**_____ with clear skin. Because this "good" strain contains genes known to fight off bacterial viruses and other potentially harmful microbes, the researchers suspect that it may actively ward off the "**6**_____" strains that are associated with disease, thereby keeping skin healthy.

"Just like good strains of bacteria in yogurt, for example, are good for the gut, these good strains of *P.* **7**_____ could be good for the skin," says Li, whose team reports the findings today in the *Journal of Investigative Dermatology*.

Acne is now often treated with antibiotics or other antimicrobial drugs. The team suggests that further studies of strain differences could lead to probiotic treatments for acne, which instead boost or supply beneficial microbes. Lotions or medications that target bad strains of bacteria or foster **8**_____ strains could offer a gentler and more effective way to ease problem skin, Li says.

"This is a great study—it was very carefully done, it addressed an important organism in the human microbiome, and it produced some very interesting results," says Martin Blaser, a physician and microbiologist at the New York University School of Medicine. He notes that the work has some limitations: It doesn't prove that the bad strains of **9**_____. *acnes* are causing **10**_____, and it doesn't explain why some people carry certain *P. acnes* strains and others don't. "But they found some strong associations," he says, "and this is a good beginning."

Swimming Pools Are Full of Poop, But They Probably Won't Make You Sick (14)

Directions: Fill in a word that best completes each sentence.

Let's start with the gross stuff: up to ten grams of poop can wash off a little kid's butt in a pool. Ten grams is a **1**_____ amount but multiply that by the number of **2**_____ in your average public pool and that's a lot of poop! Yet, most people never get an infection from swimming in a public pool. Young kids are far more likely to get sick because they end up swallowing the water, but even many of them never got seriously ill from the bacteria and parasites in swimming pools.

It's not because pools are clean. Pools are awful. They're full of poop and pee and probably some blood. Plus, lots of pools don't have the proper amount of disinfecting agents like chlorine.

There are *some* people who *do* get sick from pool water. Cryptosporidium causes about half of all outbreaks of gastrointestinal problems caught in recreational waters. Crypto one of the few microorganisms that can survive chlorine. **3**_____, when dissolved in water, breaks down into two chemicals that destroy microorganisms' protective walls. Crypto has a thick coating that doesn't allow **4**_____ to break through, so it persists even if pools are properly chlorinated. All it takes is some little kid to go swimming too soon after having an infection to spread crypto to the whole **5**_____.

If the pool doesn't have the right balance of **6**_____ chemicals, crypto isn't all you have to worry about. You can catch pus-filled rashes from Pseudomonas, a fever and cough from Legionella, or diarrhea from microbes like Shigella, Giardia, Norovirus, or E. coli.

About 8 in every 10 pool inspections find serious **7**_____ code violations. Lots of these **8**_____ have nothing to do with the number of microbes in the pool. They're more to do with whether there are enough disinfecting agents in the pool. Failing to disinfect properly is how a disease outbreak from a pool would start.

Still, there are over 300 million pool visits every year in the United States, yet only around 1,400 people get infections from the revolting **9**_____ they swim in. That's a tiny number of infections considering how filthy those pools were. More people (4,600 in 2008) got sick from pool disinfectants or died (3,300 in 2008) from drowning. Pools may be gross, but your body is great at warding off **10**_____.

Powered by Poop (15)

Directions: Fill in a word that best completes each sentence.

Every day, people poop. With more than 7 billion people on the planet, that's a lot of waste. Mixed with water, **1**_____ is called sewage. **2**_____ contains germs, so it can't be left lying around. If sewage taints the water people use for eating, drinking and bathing, those **3**_____ can spread disease. Therefore, cleaning up the daily production of human wastes is an important part of keeping people healthy.

Although we think of sewage as waste, it is also useful because it is full of energy waiting to be harnessed. That's why scientists have found ways to turn wastes into a source of renewable energy. People around the world have started using waste as a fuel source.

In England, when the GENeco Bio-Bus pulls up to a stop, everyone knows it's running on human waste: The bus is covered in tastefully drawn images of people sitting on toilets. Toilet wastes, along with discarded food, are used to create the fuel for the **4**_____. To make the fuel, waste follows a process: When toilets are flushed, or food waste is discarded, the sewage travels to the Wessex Water treatment plant. There the plant separates solid materials from the rest of the wastewater. The **5**_____ material, called sludge, is fed into an anaerobic digester. Inside this structure, bacteria digest the waste. The **6**_____ work in an oxygen-free, or anaerobic, environment. As the bacteria eat the **7**_____, they release gas, which is mostly methane. The methane gas is used as bus fuel.

Methane is a greenhouse gas, which helps trap heat close to Earth's surface; it is one cause of global warming. When methane is captured, rather than released into the air, it can be burned as a renewable source of energy. **8**_____ is a natural gas and burns cleaner than other fossil fuels like gasoline.

A wastewater treatment plant in California captures methane using a similar process. Run by the Inland Empire Utilities Agency, this plant purifies the gas and pipes it into a fuel cell. This device converts chemical energy into electric **9**_____. The fuel cell is heated before the gas comes in. When the biogas (gas collected from the breakdown of waste) enters the cell, it reacts with oxygen. This produces water, carbon dioxide and electricity.

The electricity powers the water-treatment plant. Excess heat is routed back to the anaerobic digester. There it warms the sludge, which is good because warm microbes work faster. The fuel cell cuts down on the coal or other fossil fuels needed to help run the **10**_____, says Jesse Pompa, an engineer at Inland Empire.

Probiotics Might Help Allergies, But We're Not Sure How (16)

Directions: Fill in a word that best completes each sentence.

People with seasonal allergies can treat them with antihistamine pills, nasal sprays, and eye drops. Unfortunately, sometimes those medications don't work. Another option may be probiotics.

Probiotics are microorganisms that, when ingested, may have a beneficial effect on our health. They are found in foods like yogurt and sauerkraut and now, in pills. When swallowed, they move to the colon where they join the ecosystem of billions of other bacteria known as the microbiome. Scientists have found these microbes play a role in regulating bodily functions, including that of the immune system.

The community of bacteria that live in the human gut can change, depending on factors like our environment, as well as the foods that are eaten. Certain makeups of **1**_____ are thought to provide benefits to human health, whereas others are thought to lead the way to certain diseases. The idea behind probiotics is to shift the gut microbiome to support good **2**_____ and prevent **3**_____.

Although there is evidence that suggests that intestinal bacteria play a role in the immune system, there is no evidence that any individual probiotic available on the market can reduce the severity and frequency of seasonal **4**_____. One comprehensive review of probiotics published in 2015 looked at 23 studies assessing the effectiveness of various probiotic strains on **5**_____ allergies. The review concluded that most probiotics improved allergy symptoms compared to a placebo. The problem, the researchers noted, was that all the studies used different strains of bacteria, making it impossible to make any sweeping conclusions. One study would find one kind of bacterial strain effective against grass pollen and another would find another strain effective, and yet another study would find both of those **6**_____ totally ineffective.

The evidence suggests that **7**_____ have some benefit for allergy symptoms, but scientists can't say which bacteria people should employ to treat what. "I doubt that probiotics will be good enough to replace current allergy medications anytime in the near future," says Matthew Ciorba, a gastroenterologist and director of the inflammatory bowel disease program at Washington University in Saint Louis. "If anything, these are likely to be an adjunct of therapy."

In the future, researchers need to narrow down exactly what strains provide what benefit. They also need to understand the method through which these microbes do their beneficial work. Some studies on mice show that probiotics might interfere with the way T cells (specific immune cells) function. Other animal studies suggest that the probiotics help to modify parts of the **8**_____ system called immunoglobulin E (IgE), potentially reducing their production. During an allergy attack, the immune **9**_____ produces tons of IgEs as an unnecessary, overactive response to an allergen, like pollen or grass. If scientists can understand exactly how probiotics are able to modify **10**_____, they might be able to tailor the probiotics to the allergic need.

Suggested Responses to Fill-Ins

Biochemistry

Article 1	Article 2	Article 3	Article 4	Article 5	Article 6
1-self	lipids	Proteins	menopause	phthalates	diet
2-nucleic	diseases	amino	effect	ate	diets
3-acids	comparison	acids	onset	out	whom
4-proteins	Graz	chemical	diet	cooked	fat
5-reactions	high	xenoproteins	women	eating	insulin
6-hydrophobic	lipids	bind	data	research	resistance
7-water	spectra	synthesize	survey	technique	carb
8-replicate	composition	disease	researchers	food	loss
9-RNA	Analyzer	refrigeration	age	Home	weight
10-hypothesis	Data	reactions	carbs	human	success

Article 7	Article 8	Article 9	Article 10	Article 11	Article 12	Article 13
vitamins	1-seafood	lipids	food	food	glycemic	benefits
soluble	2-less	layer	fat	crops	index	pregnant
phytochemicals	3-fertility	asymmetry	NAMPT	Puranik	low	acid
carotenoid	4-two	properties	mice	millet	carbohydrates	Consumer
form	5-servings	physical	diet	engineering	pasta	Folic
women	6-pregnant	two	mice	staple	diet	calcium
E	7-women	Levental	high	minerals	weight	supplements
supplements	8-women	outside	function	corn	cup	iron
pregnant	9-older	membrane	tissue	genetically	foods	E
people	10-history	immune	humans	modified	analysis	Vitamin

Cellular Energy

Article 1	Article 2	Article 3	Article 4	Article 5	Article 6	Article 7	Article 8
1-pit	artificial	plant	corpse	oxygen	lamp	dioxide	mistletoe
2-species	plants	insects	eat	cells	plant	sink	album
3-oxygen	sunlight	blue	pollination	ATP	nanobionics	photosynthesis	parasites
4-respiration	synthetic	flowers	female	fish	enzyme	carbon	European
5-energy	photosynthesis	light	inside	species	components	phosphorus	energy
6-mitochondria	dioxide	cells	leaf	layer	plant	fertilization	l
7-published	atmosphere	bee	years	water	solution	decreasing	Complex
8-loriciferans	discussion	petal	collapses	anaerobic	leaves	cycle	research
9-live	carbon	nectar	male	acid	light	seasonal	energy
10-life	roundtable	halo	flower	breath	light	land	nutrition

Article 9	Article 10	Article 11	Article 12	Article 13	Article 14	Article 15
1-genes	mitochondria	oxygen	genes	oxygen	Altitude	vitamins
2-mitochondria	energy	photosynthesis	functions	ancestor	high	B
3-membrane	products	simpler	mitochondria	clams	low	numbers
4-energy	respiration	anoxygenic	energy	mantle	blood	acid
5-protein	cellular	bacteria	Cell	sunlight	level	complex
6-complexes	free	evolved	respiration	algae	red	B5
7-control	radicals	diversification	proteins	photosynthesis	cell	DRI
8-independent	cell	protein	inside	zooxanthellae	train	pantothenic
9-in	efficient	a	TFB1M	slug	count	deficiency
10-Williams	fruits	tree	diseases	chloroplasts	erythropoietin	supplements

Classification

Article 1	Article 2	Article 3	Article 4	Article 5	Article 6
1-lizards	new	pouch	algae	similarities	Herpetology
2-legless	frog	species	color	structure	amphibian
3-alike	trees	nosed	multicellular	classification	reptiles
4-groups	frog	reproductive	green	system	breathe
5-snakes	cowboy	researchers	atmosphere	reproduce	eggs
6-small	catfish	hairy	bacteria	amino	characteristics
7-species	International	wombats	ancestor	conventional	neck
8-researchers	team	cycle	invaded	tools	articulation
9-pulchra	mercury	breeding	classify	descent	water
10-eyelids	human	Johnson	name	viruses	Earth

Article 7	Article 8	Article 9	Article 10	Article 11	Article 12	Article 13
1-females	highest	biomass	kingdom	challenge	group	fish
2-years	system	order	hierarchal	life	life	fish
3-faced	binomial	family	characteristics	forms	sight	diverse
4-Dog	Aristotle	diverse	root	animals	water	Ostariophysi
5-mating	scientists	backbones	systems	protists	organisms	Tree
6-two	plants	vertebrates	class	reclassification	Cryptomycota	UCEs
7-eight	botany	Birds	seed	species	fungi	Life
8-species	classification	reptile	bundles	plants	environments	relationships
9-study	Linnaeus	land	section	group	chitin	scientists
10-bat	taxonomy	Carnivores	organisms	discovered	food	relationships

Article 14	Article 15
1-Latin	name
2-system	Hoser
3-birds	taxa
4-Linnaean	paper
5-egg	reviewed
6-viviparous	peer
7-mammals	ICZN
8-milk	publish
9-protein	nomenclature
10-bacteria	Herpetologists

Ecology

Article 1	Article 2	Article 3	Article 4	Article 5	Article 6	Article 7	Article 8
1-site	ecosystem	footprint	fire	radiation	rabbits	habitat	lichens
2-phosphorus	reefs	sandwiches	ants	roaches	species	lemurs	pollution
3-13	coral	home	drown	cockroach	predator	species	air
4-sediment	nautical	carbon	worse	climate	humans	fragment	pollutant
5-dams	charts	emissions	bees	eat	dingo	isolation	bio-monitors
6-beaver	human	atmosphere	clan/group	killing	rabbits	Lemurs	quality
7-stream	coral	date	sting/bites	resistance	cane toad	loss	19th
8-water	die	shelf	humans	clean	sugarcane	species	bioindicators
9-pond	humans	waste	species	climate	species	areas	substances
10-two	climate	ingredients	fire	change	virus	metapopulations	health

Article 9	Article 10	Article 11	Article 12	Article 13	Article 14	Article 15
1-algae	cyanobacteria	population	migrate	rabbit	laughing	birds
2-lichens	microscope	endangered	distances	hamster	species	nest
3-third	symbiosis	Legal	parasites	mouths	species	range
4-fungi	partner	Wildlife	animals	pikas	extinct	hatch
5-partner	lichen	status	migrants	climate	cost	Alaska
6-cyanobacteria	diverse	manatees	hooved	rise	location	rejected
7-partner	conditions	species	infectious	temperatures	de-extinction	eggs
8-yeast	life	list	domesticated	longer	environmental	cuckoos
9-relationship	nutrients	Fish	habitats	pikas	conserve	species
10-symbiotic	landscape	threatened	harmful	change	tool	hosts

Article 16	Article 17	Article 18
1-plant	preserve	paralysis
2-leaf	menus	Mississippi
3-water	dinner	hair
4-stomata	spring	tick
5-closing	eggs	host
6-sensor	insects	symptoms
7-circuit	climate	presence
8-opening	migration	up
9-electronic	earlier	year
10-water	note	girl

Evolution

Article 1	Article 2	Article 3	Article 4	Article 5	Article 6	Article 7	Article 8
1-left	antagonism	cities	sauropsids	years	dogs	mya	tandem
2-hemisphere	strategy	evolution	wings	north	domesticated	roseae	walking
3-opposite	population	environment	flight	cycles	familiaris	fish	Dopamine
4-nerve	human	species	feathers	Earth	diverse	Tiktaalik	movements
5-perception	women	native	down	orbit	breeds	land	locomotion
6-forelimb	off	mosquitoes	insulation	magnetic	diversity	limbs	fly
7-distance	longevity	diseases	flight	ago	pure	oxygen	ISOL
8-eye	men	species	birds	sun	gene	lung	movements
9-depth	children	cities	trees	fields	domestic	swim	bird
10-vision	sexual	Urban	dinosaurs	time	dogs	tetrapods	synchronize

Article 9	Article 10	Article 11	Article 12	Article 13	Article 14	Article 15	Article 16
1-brains	island	noses	crayon	pigeons	bones	amphibians	horses
2-arteries	species	climates	ochre	words	female	land	genus
3-glucose	birds	air	pebble	star	women	predators	legged
4-species	Major	lungs	Carr	before	strain	day	toes
5-energy	mate	shapes	Star	real	Crew	bones	fossil
6-calories	Daphnie	sun	Archaeology	bigram	Team	synapsids	equine
7-calories	song	pigmentation	red	letter	rowers	receptors	francisci
8-humans	beak	D	Mesolithic	Reading	leg	heat	stilt
9-mammal	lineage	skin	life	words	loading	young	fossils
10-ancestors	genetic	adaptations	period	people	arm	hair	genus

Article 17	Article 18	Article 19
1-burgdorferi	turkeys	slender
2-diseases	England	infants
3-disease	Exeter	arms
4-borne	bones	Cobra
5-bodied	University	defense
6-life	archaeology	loris
7-Borrelia	turkeys	snake
8-tick	pottery	humans
9-iceman	wealthy	pet
10-Lyme	bones	furry

Genetics

Article 1	Article 2	Article 3	Article 4	Article 5	Article 6	Article 7	Article 8
1-Plateau	Piebaldism	chimpanzees	cars	boy	genes	size	Cacao
2-Asia	diseases	primates	weather	bacterial	social	gene	tropical
3-legend	embryo	disease	lizards	disease	friends	cells	trees
4-Tibetan	fur	research	winter	stem	people	body	Chocolate
5-yeti	skin	egg	colder	stem	friendships	mice	viruses
6-myths	front	SCNT	genes	doctors	genes	HMGA2	DNA
7-legend	animals	past	northern	grow	relationships	pig	fungi
8-Asia	pigment	technique	lizards	epidermis	extremes	copy	Cacao
9-bears	cells	large	cold	body	expressed	fetuses	pods
10-research	model	clones	die	skin	relationships	species	CRISPR

Article 9	Article 10	Article 11	Article 12	Article 13	Article 14	Article 15
1-retina	cells	flies	proteins	cells	growth	illegal
2-rods	genes	fruit	junk	organ	people	timber
3-light	gray	Hour	pluripotent	energy	activate	tree
4-melanopsin	MITF	water	neurons	off	metabolism	DNA
5-gene	interferon	Tarceva	DNA	young	nutrient	Tali
6-sensitive	hair	alcohol	genome	clock	pyruvate	illegal
7-cones	immune	cancer	intelligence	diet	stem	infrared
8-gene	innate	metabolism	HARs	aging	increased	near
9-gene	Vitiligo	Happy	genes	caloric	loss	EUR
10-therapy	life	alcoholism	protein	circadian	lactate	origin

Article 16	Article 17
1-age	heart
2-damage	stem
3-biological	genes
4-oxidative	development
5-marker	id
6-8-oxoGsn	discover
7-women	development
8-estrogen	genes
9-urine	heart
10-diseases	muscle

Human Body Systems

Article 1	Article 2	Article 3	Article 4	Article 5	Article 6	Article 7
1-study	babies	flu	hair	Hearts	breathing	dehydration
2-fecal	dogs	immune	body	muscle	cells	water
3-females	cat	levels	men	tough	rate	water
4-gastrointestinal	germs	cells	women	chicken	pre-Botzinger	body
5-children	microbiome	CD94	X	A	rate	loss
6-without	immune	symptoms	legs	organ	complex	weight
7-transplant	hypothesis	vaccine	therapy	meat	mice	pounds
8-autism	microbes	killer	Laser	women	nerve	stage
9-size	asthma	NK	waxing	Food	brain	water
10-group	life	system	bleaching	families	panic	rehydrating

Article 8	Article 9	Article 10	Article 11	Article 12	Article 13	Article 14
1-nicotine	processes	microbial	radiation	allergies	drug	honey
2-smoked	eye	fuel	sunscreens	animals	drugs	placebo
3-e-cigarettes	camera	cells	time	humans	cells	syrup
4-smoking	retina	urine	hot	food	drugs	tablespoon
5-vaping	light	battery	expiration	animals	together	day
6-smoking	surface	filaments	sun	adverse	beneficial	immune
7-survey	pliable	energy	dioxide	allergy	DNA	pollen
8-cigarettes	brain	microbial	ingredients	food	drug	birch
9-vaped	age	electrode	inactive	reactions	HIV	symptoms
10-combustible	diseases	wastes	fabric	diet	synergistic	effect

Article 15	Article 16	Article 17	Article 18	Article 19	Article 20
1-temperature	social	Acute	over	bacteria	learning
2-body	humans	term	outdoor	inflammation	another
3-degrees	oxytocin	food	allergens	IBD	academic
4-adult	hormone	pain	inside	type	strategies
5-98.6	receptor	mice	immune	syndrome	leaning
6-men	genes	inflammatory	Antihistamines	cells	students
7-day	schizophrenia	neurons	mucus	emulsifiers	techniques
8-fluctuation	chemical	Chronic	membranes	intestinal	use
9-average	cooperate	survival	corticosteroids	IBD	training
10-fever	dopamine	brain	allergist	emulsifiers	situations

Article 21	Article 22	Article 23	Article 24	Article 25	Article 26
1-activity	day	Takebe	reactions	Stool	systems
2-physical	subjects	cells	gene	intestine	immune
3-benefits	lipids	cells	allergic	large	mice
4-sedentary	tissue	iPS	peanut	food	cells
5-weight	muscle	HUVECs	ingested	tract	infections
6-calorie	subjects	pancreatic	after	Constipation	mice
7-obesity	circadian	islets	blood	Western	suppression
8-calories	insulin	transplant	food	form	babies
9-resistance	bilayer	diabetes	peanut	style	CD71
10-control	diabetes	vascularize	study	fiber	prematurely

Article 27	Article 28	Article 29	Article 30	Article 31	Article 32
1-salt	questions	device	stiff	balloon	vitamin
2-table	self	fiber	stiffness	volume	D
3-taste	themselves	bandage	thing	pressure	exposure
4-diet	Children	medications	back	relationship	sun
5-sodium	parents	bandage	measured	inputs	equator
6-blood	self-esteem	cost	feeling	eating	cancers
7-water	growth	battlefield	measurement	stomach	north
8-urination	mindset	bandage	people	food	Melanoma
9-diets	children	experiment	pain	exit	UV
10-retaurants	self-concept	testing	clinic	brain	UVA

Article 33	Article 34	Article 35	Article 36
1-surgery	suicide	avoid	sleep
2-condition	violence	system	problems
3-neurologic	years	behavioral	children
4-brain	try	sources	ages
5-surgery	patients	disgust	factors
6-Aquarium	illness	Curtis	risk
7-shunt	mental	Barra	sleep
8-abdomen	men	disease	problems
9-Ziggy	women	categories	life
10-Mystic	suicide	-environment	older

Reproduction & Development

Article 1	Article 2	Article 3	Article 4	Article 5	Article 6	Article 7	Article 8
1-family	life	colors	cells	tail	superfecundation	mammals	repeats
2-sisters	eat	butterfly	human	regenerate	zygote	egg	telomere
3-chromosome	mouse	cameras	bud	geckos	heteropaternal	uterine	clock
4-sperm	brain	cocoons	tubercle	cords	cycle	proteins	telomerase
5-gene	leptin	cameras	tail	lizard	egg	placental	molecular
6-daughters	weight	butterflies	phalluses	predator	menstrual	inflammatory	telomere
7-son	brain	scale	limb	stem	fathers	17	synthesis
8-men	gain	scale	genes	rest	queens	uterus	signal
9-war	plastic	cocoon	Genital	scar	tomcats	interleukin	stem
10-boys	babies	butterfly	snakes	tissue	kittens	response	cancer

Article 9	Article 10	Article 11	Article 12	Article 13
1-cells	bone	water	males	place
2-ball	alcohol	fertilizing	female	mice
3-implantation	cancer	Self	dragons	skin
4-placenta	drug	killifish	eggs	injuries
5-women	protein	nucleotides	reversed	healed
6-Pregnancies	disulfiram	nucleotides	temperature	injury
7-NK	patients	fish	sex	genes
8-cells	spreading	mutations	determination	DNA
9-mice	drug	lineage	bearded	stem
10-miscarriage	trials	stranger	evolutionary	cells

Scientific Inquiry

Article 1	Article 2	Article 3	Article 4	Article 5	Article 6
1-school	caffeinated	cat	blood	light	dogs
2-life	gum	causation	Cretaceous	task	expressions
3-high	athletes	popsicles	tick	rodents	facial
4-later	coffee	correlation	feathered	bright	attention
5-behaviors	sprints	cancer	flight	rats	person
6-reading	caffeine	psychosis	dinosaur	dim	muscles
7-attainmnet	consumption	ownership	amber	hippocampus	humans
8-parental	sports	studies	Burmese	brain	communicate
9-cognitive	caffeine	T. Gondii	dinosaur	bright	individual's
10-traits	performance	feces	blood	neurons	humans

Article 7	Article 8	Article 9	Article 10	Article 11	Article 12	Article 13
1-driving	chimpanzees	distinct	moderate	helpers	video	traits
2-mind	research	individual	people	chimpanzees	game	music
3-do	Haven	shrew	weight	bonobos	study	brain
4-wander	sanctuary	box	radiologist	Man	realism	music
5-environment	NIH	personality	weight	helpful	world	tastes
6-car	species	bees	empty	apple	physics	systemizer
7-driving	monkeys	decisions	brain	unhelpful	game	empathizers
8-environment	neighbors	species	obesity	humans	violent	musical
9-urban	Suomi's	impulsive	procedure	dominant	Zendle	taste
10-phone	announcement	variety	size	preference	video	type

Article 14	Article 15	Article 16	Article 17	Article 18
1-disgusted	jet	twins	happy	brains
2-people	activity	facial	social	neurons
3-smells	owls	identical	media	carnivores
4-groups	academic	recognition	time	hypothesis
5-disease	academic	face	screen	herbivores
6-scale	lagged	future	digital	carnivores
7-odor	students	twins	hours	predators
8-Donald	larks	lives	happiness	energy
9-political	finches	genes	teens	neurons
10-changed	learning	environment	smartphone	brains

Study of Life

Article 1	Article 2	Article 3	Article 4	Article 5	Article 6
1-reproduction	bacteria	study	rabbits	RNA	transplants
2-life	toxins	bacteria	sick	negative	Fecal
3-viruses	strains	families	symptoms	bacteria	Clostridium
4-RNA	colony	children	bacterium	disease	transplantation
5-Tree	attack	positive	medication	Gram	Bacteria
6-information	disease	illness	dog	nucleatum	gram
7-cell	toxins	foodborne	rabbits	F	Bordenstein
8-evolution	warfare	coli	bacteria	salivary	cases
9-virologists	gut	kitchen	rabbit	antibiotic	bacterial
10-branches	provoke	illness	fever	positive	components

Article 7	Article 8	Article 9	Article 10	Article 11	Article 12
1-poop	bacteria	gut	compound	T	division
2-fecal	germs	bacteria	nose	aptata	pneumonia
3-pills	sick	mummies	bacterium	Habelia	stain
4-material	wash	Egypt	S. aureus	arthropods	layer
5-microbiome	surface	human	antibiotic	Chelicerae	less
6-intestines	phone	Akkermansia	S. lugdunensis	appendages	environment
7-difficile	time	muciniphilia	microbial	mandibulates	cell
8-transplant	clean	Unhealthy	drugs	appendages	water
9-bacteria	cleaning	barrier	years	chelicerates	peptidoglycan
10-body	mouth	bacteria	lugdunin	Explosion	positive

Article 13	Article 14	Article 15	Article 16
1-bacteria	small	waste	bacteria
2-skin	kids	Sewage	health
3-clear	Crypto	germs	diseases
4-strains	chlorine	bus	allergies
5-people	pool	solid	seasonal
6-bad	disinfecting	bacteria	strains
7-acnes	health	sludge	probiotics
8-good	violations	Methane	immune
9-P	pools	energy	system
10-acne	infections	plant	IgEs

Sample Exit Tickets

Each fill-in/warm-up page has a blank lined page that follows it. Students can complete the warm-up at the start of class and save it till the end of class when they are given "Exit Ticket Statement Starters", which can be displayed for students to copy and complete on the back of the warm-up. Students might complete two-three statements in the last five minutes of class before they dismiss. Here are some samples, which can be rotated throughout the school year:

The main idea of today's class was….

The concept that I'm most confused about from today's class is….

The thing that I need to practice most from today's topic is…

Today's lesson can be used in the "real world" by….

I could summarize today's lesson by saying…

Today I learned….

One question I still have about today's lesson is…

The thing I found most interesting about today's lesson is….

One way I could apply what we learned in class today to my everyday life is….

The essential concept we explored today in class was…

Two similarities about ideas we learned today are…

Two differences about ideas we learned today are…

Write two sentences using two vocabulary words from today's lesson.

The most important point I took away from today's class was…

I am confused about _____ from today's lesson.

I'd like to hear more about….

What we learned today is useful because….

What we learned today fits into the main unit topic in the following way…

An idea from today's class that is relevant to the world is….

I predict that the topic of tomorrow's class will be…

Elaborate on a concept from today's lesson.

Explain how the process we learned about in class today works.

The thing that surprised me most about today's class was…

In our next class I'd like to review…

In today's class I accomplished….

Bibliography

Biochemistry

Ballantyne, C. (2007). Fact or fiction? Vitamin supplements improve your health. *Scientific American.* Retrieved from https://www.scientificamerican.com/article/fact-or-fiction-vitamin-supplements-improve-health/

Biophysical Society. (2018, February 20). 'Lipid asymmetry' plays key role in activating immune cells. *ScienceDaily.* Retrieved August 21, 2018 from www.sciencedaily.com/releases/2018/02/180220083847.htm

Consumer Reports. (2017) *The Washington Post.* Retrieved from https://www.washingtonpost.com/national/health-science/some-vitamins-and-minerals-may-carry-more-risks-than-benefits/2017/04/14/01ed4166-c2f0-11e6-9a51-cd56ea1c2bb7_story.html?noredirect=on&utm_term=.ffe27a35117e

Cepelewicz, J. (2017) Life's first molecule was protein, not RNA, new model suggests. *Quanta Magazine.* Retrieved from https://www.scientificamerican.com/article/life-rsquo-s-first-molecule-was-protein-not-rna-new-model-suggests/

Geggel. L. (2018). Can your diet delay menopause? *Live Science.* Retrieved from https://www.livescience.com/62443-eating-peas-fish-may-delay-menopause.html

George Washington University. (2018, March 29). Dining out associated with increased exposure to harmful chemicals: New study finds burgers and other foods consumed at restaurants, fast food outlets or cafeterias, associated with higher levels of phthalates. *ScienceDaily.* Retrieved August 23, 2018 from www.sciencedaily.com/releases/2018/03/180329095722.htm

Graz University of Technology. (2017, October 23). Boost for lipid research: Researchers facilitate lipid data analysis. *ScienceDaily.* Retrieved August 19, 2018 from www.sciencedaily.com/releases/2017/10/171023123516.htm

Merrifield, Rex. (2017). New strains of staple crops serve up essential vitamins. *Horizon.* Retrieved from https://horizon-magazine.eu/article/new-strains-staple-crops-serve-essential-vitamins_en.html

Nierenberg. C. (2018). Could eating more seafood help couples conceive? *Live Science.* Retrieved from https://www.livescience.com/62660-seafood-diet-fertility.html

Nierenberg, C. (2018). Why you probably shouldn't waste your money on DNA-Based diets. *Live Science.* Retrieved from https://www.livescience.com/61807-do-dna-diets-work.html

Rettner, R. (2018). Can eating pasta really help you lose weight? *Live Science.* Retrieved from https://www.livescience.com/62219-pasta-weight-loss.html

Trafton, A. (2018). Chemists synthesize millions of proteins not found in nature. *MIT News.* Retrieved from http://news.mit.edu/2018/chemists-synthesize-millions-proteins-not-found-nature-0521

University of Copenhagen The Faculty of Health and Medical Sciences. (2018, May 3). Researchers defy biology: Mice remain slim on burger diet. *ScienceDaily*. Retrieved June 20, 2018 from www.sciencedaily.com/releases/2018/05/180503142706.htm

Cellular Energy

Bradford, A. (2017), Corpse flower: Facts about the smelly plant. *Live Science*. Retrieved from https://www.livescience.com/51947-corpse-flower-facts-about-the-smelly-plant.html

Bradford, A. (2018). What is vitamin B5 (Pantothenic acid)? *Live Science*. Retrieved from https://www.livescience.com/51640-b5-pantothenic-acid.html

Davis, N. (2017). Flowers use 'blue halo' optical trick to attract bees, say researchers. *The Guardian.* Retrieved from https://www.theguardian.com/science/2017/oct/18/flower-nanostructures-optical-trick-attract-bees-pollinators-blue-halo

Engber, D. (2015). Do fish get out of breath? *Popular Science*. Retrieved from https://www.popsci.com/do-fish-get-out-breath

Fox, S. (2010). First ever multicellular animals found In oxygen-free environment. *Popular Science*. Retrieved from https://www.popsci.com/science/article/2010-04/first-ever-multicellular-animals-found-oxygen-free-environment

Hamers, L. (2016) Why do our cell's power plants have their own DNA? *Science, 361*(6404), doi:10.1126/science.aaf4083

Imperial College London. (2016, March 15). Photosynthesis more ancient than thought, and most living things could do it. *ScienceDaily*. Retrieved June 19, 2018 from www.sciencedaily.com/releases/2016/03/160315104148.htm

Karolinska Institute. (2009, April 14). Key protein in cellular respiration discovered. *ScienceDaily*. Retrieved June 20, 2018 from www.sciencedaily.com/releases/2009/04/090408074426.htm

Lecher, C. (2012). FYI: Does training at high altitudes help olympians win? *Popular Science*. Retrieved from https://www.popsci.com/science/article/2012-08/fyi-does-training-high-altitudes-help-olympians-win

McBried, H. Neuspiel, M. Wasiak, S. (2006). Mitochondria: More than just a powerhouse. *Current Biology. 16*(14), R551-R560. https://doi.org/10.1016/j.cub.2006.06.054

Ornes, S. (2010). The algae invasion. *Science News for Students.* Retrieved from https://www.sciencenewsforstudents.org/article/algae-invasion

The Kavli Foundation. (2015, September 8). Artificial 'plants' could fuel the future: By combining semiconducting nanowires and bacteria, researchers can now produce liquid fuel. *ScienceDaily*. Retrieved June 19, 2018 from www.sciencedaily.com/releases/2015/09/150908144311.htm

Trafton, A. (2017). Engineers create plants that glow. Anne Trafton. *MIT News*. Retrieved from http://news.mit.edu/2017/engineers-create-nanobionic-plants-that-glow-1213

University of Exeter. (2016, October 3). Future increase in plant photosynthesis revealed by seasonal carbon dioxide cycle. *ScienceDaily*. Retrieved June 20, 2018 from www.sciencedaily.com/releases/2016/10/161003112208.htm

Williams, S. (2018). Mistletoe lacks key energy-generating complex. *The Scientist*. Retrieved from https://www.the-scientist.com/?articles.view/articleNo/52488/title/Mistletoe-Lacks-Key-Energy-Generating-Complex/

Classification

Boyle, R. (2012). Rainforest expedition turns up 46 new creatures, including this cowboy frog. *Popular Science*. Retrieved from https://www.popsci.com/science/article/2012-01/rainforest-expedition-turns-46-new-creatures-including-cowboy-frog

Diep, F. (2013). 4 New legless lizards discovered in California. *Popular Science*. Retrieved from https://www.popsci.com/science/article/2013-09/four-new-legless-lizards-discovered-california

Dillow, C. (2011). Biologists announce discovery of an entirely new branch of life. *Popular Science*. Retrieved from https://www.popsci.com/science/article/2011-05/biologists-discover-entirely-new-branch-life-living-microscopic-level

Dockrill, P. (2018). The milk of australia's weirdest animal could help us fight antibiotic resistance. *Science Alert*. Retrieved from https://www.sciencealert.com/this-quirky-australian-critter-best-hope-against-antibiotic-resistance-platypus-antimicrobial-milk

Jones, B. (2017). A few bad scientists are threatening to topple taxonomy. *Smithsonian*. Retrieved from https://www.smithsonianmag.com/science-nature/the-big-ugly-problem-heart-of-taxonomy-180964629/

Louisiana State University. (2017, November 3). What do piranhas and goldfish have in common? *ScienceDaily*. Retrieved June 12, 2018 from www.sciencedaily.com/releases/2017/11/171103105632.htm

Nobel, S. (2017). Differences between reptiles and amphibians. *Reptile Blog*. Retrieved from http://www.reptilesmagazine.com/Differences-Between-Reptiles-and-Amphibians/

Ornes, S. (2010). The algae invasion. *Science News for Students*. Retrieved from https://www.sciencenewsforstudents.org/article/algae-invasion

Sharp, J. (n.d.). How are animals classified? *Desert USA*. Retrieved from https://www.desertusa.com/desert-activity/classified-plants-animals.html

Sharp, J. (n.d.). How are plants classified? *Desert USA*. Retrieved from https://www.desertusa.com/flora/plant-classified.html

Sinclair RM, Ravantti JJ, Bamford DH. 2017. Nucleic and amino acid sequences support structure-based viral classification. *Journal of Virology*, 91:e02275-16. https://doi.org/10.1128/JVI.02275-16.

Tilton, L. (2009). From Aristotle to Linnaeus: the History of Taxonomy. *Dave's Garden*. Retrieved from https://davesgarden.com/guides/articles/view/2051

Two new dog-faced bat species discovered. (2018). *Science News.* Retrieved from http://www.sci-news.com/biology/two-new-dog-faced-bat-species-05828.html

University of Queensland. (2018, May 9). Breeding benefits when love bites wombats on the butt. *ScienceDaily.* Retrieved June 20, 2018 from www.sciencedaily.com/releases/2018/05/180509104956.htm

Wiley. (2012, September 27). Nature's misfits: Reclassifying protists helps answer how many species remain undiscovered. *ScienceDaily.* Retrieved June 20, 2018 from www.sciencedaily.com/releases/2012/09/120927124202.htm

Ecology

Bilba, E. (2017). Inside Australia's war on invasive species. *Scientific American.* Retrieved from https://www.scientificamerican.com/article/inside-australia-rsquo-s-war-on-invasive-species/

Callaghan, M. (2016). Lichens can be made of three organisms, not just two. *The Naked Scientists.* Retrieved from https://www.popsci.com/new-research-finds-lichens-are-not-just-two-organism-marriage

Codosh, S. (2017). Taking manatees off the endangered species list doesn't mean we should stop protecting them. *Popular Science.* Retrieved from https://www.popsci.com/taking-manatees-off-endangered-species-list-doesnt-mean-well-stop-protecting-them

Eschner, K. (2018). Russian cuckoos are taking over Alaska. *Popular Science.* Retrieved from https://www.popsci.com/russian-cuckoo-invading-alaska

Fecht, S. (2017). Resurrecting extinct animals might do more harm than good. *Popular Science.* Retrieved from https://www.popsci.com/resurrecting-extinct-animals-might-harm-more-species-than-it-helps

Intagliata, C. (2017). Rabbit relatives reel from climate change. *Scientific American.* Retrieved from https://www.scientificamerican.com/podcast/episode/rabbit-relatives-reel-from-climate-change/

Mikulec, M. (2018). Hunting down air pollution: Meet the lichens. *The Naked Scientists.* Retrieved from https://www.thenakedscientists.com/articles/science-features/hunting-down-air-pollution-meet-lichens

PLOS. (2018, May 9). For lemurs, size of forest fragments may be more important than degree of isolation: Occurrence of these endangered primates rises with patch size but is mixed for patch connectivity. *ScienceDaily.* Retrieved June 20, 2018 from www.sciencedaily.com/releases/2018/05/180509162709.htm

Rosen, J. (2017). Springtime now arrives earlier for birds. *Scientific American.* Retrieved from https://www.scientificamerican.com/podcast/episode/springtime-now-arrives-earlier-for-birds/

Schweid, R. (2016). If people were cockroaches, adapting to climate change would be easy. *Popular Science.* Retrieved from https://www.popsci.com/if-people-were-cockroaches-adapting-to-climate-change-would-be-easy

The terrifying way fire ants take advantage of hurricane floods. Ellen Airhart August 29, 2017. https://www.popsci.com/fire-ants-take-advantage-hurricane-harvey-floods

Trafton, A. (2017). Sensors applied to plant leaves warn of water shortage. *MIT News.* Retrieved from http://news.mit.edu/2017/sensors-applied-plant-leaves-warn-water-shortage-1108

University of Exeter. (2018, May 9). Beavers do 'dam' good work cleaning water. *ScienceDaily.* Retrieved June 20, 2018 from www.sciencedaily.com/releases/2018/05/180509121552.htm

University of Georgia. (2018, May 8). Migratory animals carry more parasites. *ScienceDaily.* Retrieved June 20, 2018 from www.sciencedaily.com/releases/2018/05/180508150321.htm

University of Manchester. (2018, January 25). What is the environmental impact of your lunch-time sandwich? *ScienceDaily.* Retrieved June 19, 2018 from www.sciencedaily.com/releases/2018/01/180125085116.htm

Weeler, M. (2017). Old maps highlight new understanding of coral reef loss. *The Naked Scientists.* Retrieved from https://www.thenakedscientists.com/articles/science-news/old-maps-highlight-new-understanding-coral-reef-loss

Wootson, C. (2018). A 5-year-old girl's sudden paralysis was a mystery. *The Washington Post.* Retrieved from https://www.washingtonpost.com/news/to-your-health/wp/2018/06/11/a-5-year-old-girls-sudden-paralysis-was-a-mystery-then-her-mother-checked-her-scalp/?utm_term=.f854c4c291d1

Evolution

Best, S. (2017). Leftovers of England's first Christmas dinner? Archaeologists discover three 16th century turkey bones under a street in Exeter. *Daily Mail.* Retrieved from http://www.dailymail.co.uk/sciencetech/article-5195147/Is-evidence-UKs-Christmas-dinner.html

Bryant, C. (n.d.) How did language evolve? *How Stuff Works.* Retrieved from https://science.howstuffworks.com/life/evolution/language-evolve.htm

Cole, S. (2016). Pigeons can read a LIttle bit, new research shows. *Popular Science.* Retrieved from https://www.popsci.com/pigeons-reading-study

Gore, R. (n.d.) The rise of mammals. *National Geographic.* Retrieved from https://www.nationalgeographic.com/science/prehistoric-world/rise-mammals/

Larsson, M. (2017). How did 3D vision develop? *The Naked Scientists.* Retrieved from https://www.thenakedscientists.com/articles/features/how-did-3d-vision-develop

Massy-Beresford, H. (2017). The battle of the sexes can show us how to live longer, say researchers. *Horizon.* Retrieved from https://horizon-magazine.eu/article/battle-sexes-can-show-us-how-live-longer-say-researchers_en.html

Milliken, G. (2015). The first dogs may have been domesticated In central Asia. *Popular Science.* Retrieved from https://www.popsci.com/dogs-may-have-been-first-domesticated-in-central-asia

Princeton University. (2017, November 24). New species can develop in as little as two generations, Galapagos study finds. *ScienceDaily.* Retrieved June 20, 2018 from www.sciencedaily.com/releases/2017/11/171124084320.htm

Rutgers University. (2018, May 7). Earth's orbital changes have influenced climate, life forms for at least 215 million years: Gravity of Jupiter and Venus elongates Earth's orbit every 405,000 years, Rutgers-led study confirms. *ScienceDaily*. Retrieved June 20, 2018 from www.sciencedaily.com/releases/2018/05/180507153109.htm

Smith, R. (2017). Humans don't use as much brainpower as we like to think. *Duke Today.* Retrieved from https://today.duke.edu/2017/10/humans-dont-use-much-brainpower-we-think

Stallard, B. (2014). Lyme disease is older than humanity. *Nature World News.* Retrieved from https://www.natureworldnews.com/articles/7304/20140529/lyme-disease-older-humanity.htm

Stephens, T. (2017) A horse is a horse, of course, of course—except when it isn't. *UC Santa Cruz New Center.* Retrieved from https://news.ucsc.edu/2017/11/ancient-horse.html

Thomson, L. (2017) Tales of evolution: How fish conquered the land. *The Naked Scientists.* Retrieved from https://www.thenakedscientists.com/articles/features/tales-evolution-how-fish-conquered-land

Thomson, L. (2018). Tales of evolution: When dinosaurs took flight. *The Naked Scientists.* Retrieved from https://www.thenakedscientists.com/articles/science-features/tales-evolution-when-dinosaurs-took-flight

University of Cambridge. (2017, November 29). Prehistoric women had stronger arms than today's elite rowing crews. *ScienceDaily*. Retrieved June 17, 2018 from www.sciencedaily.com/releases/2017/11/171129143359.htm

University of Toronto. (2017, November 2). Are cities affecting evolution? Explosion of rats, clovers, bedbugs, mosquitoes unintended evolutionary consequence of urbanization. *ScienceDaily*. Retrieved June 20, 2018 from www.sciencedaily.com/releases/2017/11/171102180520.htm

University of York. (2018, January 26). Archaeologists say they may have discovered one of the earliest examples of a 'crayon'. *ScienceDaily.* Retrieved June 19, 2018 from www.sciencedaily.com/releases/2018/01/180126095323.htm

Yin, S. (2017). Ancestral climates may have shaped your nose. *The New York Times.* Retrieved from https://www.nytimes.com/2017/03/16/science/ancestral-climates-may-have-shaped-your-nose.html

Zhu. A. (2014). Venomous slow loris may have evolved to mimic cobras. *Popular Science.* Retrieved from https://www.popsci.com/article/science/venomous-slow-loris-may-have-evolved-mimic-cobras.

Genetics

Bogt-James, M. (2017). UCLA scientists identify a new way to activate stem cells to make hair grow. *UCLA Newsroom.* Retrieved from http://newsroom.ucla.edu/releases/ucla-scientists-identify-a-new-way-to-activate-stem-cells-to-make-hair-grow

Bryon-Dodd, K. (2017). Gene therapy restores vision in blind mice. *BioNews.* Retrieved from https://www.bionews.org.uk/page_96207

Callaway, E. (2009). 'Happyhour' gene may help put boozers off their drink. *NewScientist.* Retrieved from https://www.newscientist.com/article/dn17178-happyhour-gene-may-help-put-boozers-off-their-drink/

Frontiers. (2018, February 27). Simple urine test could measure how much our body has aged: A promising new marker of aging could help predict the risk of developing age-related disease and even death. *ScienceDaily*. Retrieved June 20, 2018 from www.sciencedaily.com/releases/2018/02/180227090733.htm

Geggel, L. (2018). Can gene editing save the world's chocolate? *Scientific American.* Retrieved from https://www.scientificamerican.com/article/can-gene-editing-save-the-worlds-chocolate/

Gonzalez, J. (2017). Tree rings used to counter smugglers' rings. *Horizon.* Retrieved from https://horizon-magazine.eu/article/tree-rings-used-counter-smugglers-rings_en.html

Intagliata, C. (2017) Cold snap shapes lizard survivors. *Scientific American.* Retrieved from https://www.scientificamerican.com/podcast/episode/cold-snap-shapes-lizard-survivors/

Maron, D. (2018). First primate clones produced using the "Dolly" method. *Scientific American.* Retrieved from https://www.scientificamerican.com/article/first-primate-clones-produced-using-the-ldquo-dolly-rdquo-method/

National University of Singapore. (2017, August 22). You may be as friendly as your genes. *ScienceDaily*. Retrieved June 20, 2018 from www.sciencedaily.com/releases/2017/08/170822103053.htm

North Carolina State University. (2018, May 7). Researchers find genetic 'dial' can control body size in pigs. *ScienceDaily*. Retrieved June 20, 2018 from www.sciencedaily.com/releases/2018/05/180507174015.htm

PLOS. (2018, May 3). Gray hair linked to immune system activity and viral infection: Researchers report that loss of hair pigmentation, or gray hair, is associated with activation of the innate immune system in mice. *ScienceDaily*. Retrieved June 20, 2018 from www.sciencedaily.com/releases/2018/05/180503142934.htm

Rice, D. (2017). Abominable snowman? Nope: Study ties DNA samples from infamous Yetis to Asian bears. *USA TODAY*. Retrieved from https://www.usatoday.com/story/tech/science/2017/11/29/abominable-snowman-nope-study-ties-dna-samples-infamous-yetis-asian-bears/902632001/

Sanford-Burnham Prebys Medical Discovery Institute. (2017, August 22). Where do heart cells come from? Id genes play surprise role in cardiac development. *ScienceDaily*. Retrieved June 18, 2018 from www.sciencedaily.com/releases/2017/08/170822104843.htm

Smith, C. (2017). Doctors fix genetic defect and grow boy new skin. *The Naked Scientists.* Retrieved from https://www.thenakedscientists.com/articles/science-news/doctors-fix-genetic-defect-and-grow-boy-new-skin

University of California, Irvine. (2017, August 10). Link between biological clock and aging revealed. *ScienceDaily.* Retrieved June 19, 2018 from www.sciencedaily.com/releases/2017/08/170810082416.htm

Yates, C. (2016). Why some cats look like they are wearing tuxedos. *Popular Science.* Retrieved from https://www.popsci.com/why-some-cats-look-like-they-are-wearing-tuxedos

Zorich, Z. (2018). Is "Junk DNA" what makes humans unique? *Scientific American.* Retrieved from https://www.scientificamerican.com/article/is-junk-dna-what-makes-humans-unique/

Human Body Systems

American Society for Biochemistry and Molecular Biology (ASBMB). (2017, October 2). When HIV drugs don't cooperate. *ScienceDaily.* Retrieved April 23, 2018 from www.sciencedaily.com/releases/2017/10/171002161730.htm

Brookhaven National Laboratory (2008). More sun exposure may be good for some people. Retrieved from https://www.bnl.gov/newsroom/news.php?a=110726

Cincinnati Children's Hospital Medical Center. (2018, May 8). Tissue-engineered human pancreatic cells successfully treat diabetic mice: Self-condensation process for cells generates vascularized organ tissues for transplant. *ScienceDaily.* Retrieved June 20, 2018 from www.sciencedaily.com/releases/2018/05/180508111745.htm

Chodosh, S. (2017). Celebrate valentine's day by eating an actual heart (seriously) *Popular Science.* Retrieved from https://www.popsci.com/organ-meat-heart-nutritious-eat

Chodosh, S. (2018). Local honey might help your allergies—but only if you believe it will. *Popular Science.* Retrieved from https://www.popsci.com/local-honey-allergies

Chodosh, S. (2018). What's the difference between indoor and outdoor allergies? *Popular Science.* Retrieved from https://www.popsci.com/indoor-and-outdoor-allergies

Collins, C. (2017). Human-dog bond provides clue to treating social disorders. *Horizon.* Retrieved from https://horizon-magazine.eu/article/human-dog-bond-provides-clue-treating-social-disorders_en.html

Couzin-Frankel, J. (2013). A new reason why newborns can't fight colds. *Science.* Retrieved from http://www.sciencemag.org/news/2013/11/new-reason-why-newborns-cant-fight-colds

DeNoon, D. (n.d.) Many babies healthier in homes with dogs. *WebMD.* Retrieved from https://www.webmd.com/parenting/baby/news/20120709/many-babies-healthier-in-homes-with-dogs#1

Eye problems: What to expect as you age. (n.d.) in *WebMD.* Retrieved from https://www.webmd.com/eye-health/vision-problems-aging-adults#1

Faerman, Z. (2017). What your poop says about your health. *Self.* Retrieved from https://www.self.com/story/what-your-poop-says-about-your-health

Frontiers. (2017, September 28). Students know about learning strategies -- but don't use them: Study suggests university students need training on how and when to apply self-regulated learning strategies for specific learning situations. *ScienceDaily.* Retrieved June 20, 2018 from www.sciencedaily.com/releases/2017/09/170928142106.htm

James Cook University. (2017, September 28). Answer to young people's persistent sleep problems. *ScienceDaily.* Retrieved June 15, 2018 from www.sciencedaily.com/releases/2017/09/170928094158.htm

Kaplan, S., (2018). E-Cigarettes' risks and benefits: Highlights from the report to the F.D.A. *New York Times.* Retrieved from https://www.nytimes.com/2018/01/23/health/e-cigarettes-health-evidence.html

Maldarelli, C. (2017). This is what happens to your body as you die of dehydration. *Popular Science.* Retrieved from https://www.popsci.com/dehydration-death-thirst-water

Maldarelli, C. (2018). How to tell if you really have a fever. *Popular Science.* Retrieved from https://www.popsci.com/check-for-fever

Mount Sinai Health System. (2017, December 5). Six genes driving peanut allergy reactions identified. *ScienceDaily.* Retrieved June 20, 2018 from www.sciencedaily.com/releases/2017/12/171205092140.htm

Ohio State University. (2017, January 23). Autism symptoms improve after fecal transplant, small study finds: Parents report fewer behavioral and gastrointestinal problems; gut microbiome changes. *ScienceDaily.* Retrieved June 20, 2018 from www.sciencedaily.com/releases/2017/01/170123094638.htm

Park, A. (2018). A better flu shot may be on the horizon. *Time.* Retrieved from http://time.com/5309937/flu-blood-test/

Servick, K. (2015). Common ingredient in packaged food may trigger inflammatory disease. *Science.* Retrieved from http://www.sciencemag.org/news/2015/02/common-ingredient-packaged-food-may-trigger-inflammatory-disease

Smith, C. (2017). Take a deep breath. *The Naked Scientists.* Retrieved from https://www.thenakedscientists.com/articles/science-news/take-deep-breath

University of Alberta. (2017, September 26). Does your back feel stiff? Well, it may not actually be stiff, study finds. *ScienceDaily.* Retrieved June 20, 2018 from www.sciencedaily.com/releases/2017/09/170926112015.htm

University of Amsterdam. (2017, September 28). Self-esteem in kids: Lavish praise is not the answer, warmth is. *ScienceDaily.* Retrieved June 20, 2018 from www.sciencedaily.com/releases/2017/09/170928085101.htm

University of Bath. (2016, April 18). Urine turned into sustainable power source for electronic devices. *ScienceDaily.* Retrieved June 20, 2018 from www.sciencedaily.com/releases/2016/04/160418095918.htm

University of Geneva. (2017, October 2). Our muscles measure the time of day. *ScienceDaily*. Retrieved June 19, 2018 from www.sciencedaily.com/releases/2017/10/171002084843.htm

University of Pennsylvania. (2018, March 22). Being hungry shuts off perception of chronic pain. *ScienceDaily*. Retrieved June 20, 2018 from www.sciencedaily.com/releases/2018/03/180322125024.htm

University of Veterinary Medicine -- Vienna. (2017, August 23). Comparing food allergies: Animals and humans may have more in common than you think. *ScienceDaily*. Retrieved June 20, 2018 from www.sciencedaily.com/releases/2017/08/170823094121.htm

Vance, E. (2018). Why are suicide rates rising? *Live Science*. Retrieved from https://www.livescience.com/62781-why-are-suicide-rates-rising.html

Weisberger, M. (2018). Is expired sunscreen better than no sunscreen? *Live Science*. Retrieved from https://www.livescience.com/62783-does-expired-sunscreen-work.html

Widdows, H. (2018). Body hair is natural. Society thinking otherwise Is dangerous. *Time*. Retrieved from http://time.com/5300646/body-hair-shaving-waxing-ethics-beauty/

Reproduction and Growth

Arizona State University. (2018, February 27). Hidden secret of immortality enzyme telomerase: Can we stay young forever, or even recapture lost youth? *ScienceDaily*. Retrieved June 20, 2018 from www.sciencedaily.com/releases/2018/02/180227142114.htm

Carrol, M. (2016). How can twins have two different fathers? *Independent*. Retrieved from https://www.independent.co.uk/life-style/health-and-families/health-news/how-can-twins-have-two-different-fathers-a6928696.html

Diep, F. (2014). Where do genitals come from? *Popular Science*. Retrieved from https://www.popsci.com/article/science/where-do-genitals-come

Griggs, M. (2017). Scientists are puzzling out how butterflies assemble their brightly colored scales. *Popular Science*. Retrieved from https://www.popsci.com/butterfly-scales

Kaiser, J. (2017). An old drug for alcoholism finds new life as cancer treatment. *Science*. Retrieved from http://www.sciencemag.org/news/2017/12/old-drug-alcoholism-finds-new-life-cancer-treatment

Maxmen, A. (2018). Armadillo and rabbit genes reveal how pregnancy evolved. *Nature*. Retrieved from https://www.nature.com/articles/d41586-018-00341-w

Milliken, G. (2016). Sex reversal in bearded dragons creates females that behave like males. *Popular Science*. Retrieved from https://www.popsci.com/sex-reversal-in-bearded-dragons-creates-females-that-behave-like-males

Newcastle University. (2008, December 12). Boy or girl? It's in the father's genes. *ScienceDaily*. Retrieved June 20, 2018 from www.sciencedaily.com/releases/2008/12/081211121835.htm

PhysOrg. (2017). Self-fertilizing fish have surprising amount of genetic diversity. Retrieved from https://phys.org/news/2017-12-self-fertilizing-fish-amount-genetic-diversity.html

Smith, C. (2017). Skin cells remember previous injury. *The Naked Scientists.* Retrieved from https://www.thenakedscientists.com/articles/science-news/skin-cells-remember-previous-injury

Thomson, L. (2017). How mum's immune cells help her baby grow. *The Naked Scientist.* Retrieved from https://www.thenakedscientists.com/articles/science-news/how-mums-immune-cells-help-her-baby-grow

University of Guelph. (2017, November 2). Cells driving gecko's ability to re-grow its tail identified: Discovery of which cells are behind the gecko's ability to re-grow its tail has implications for spinal cord treatment in humans. *ScienceDaily.* Retrieved June 20, 2018 from www.sciencedaily.com/releases/2017/11/171102120954.htm

Zhang, C. (2017). Our mothers' exposure to BPA might lead us to binge as adults. *Popular Science.* Retrieved from https://www.popsci.com/our-mothers-exposure-to-bpa-might-lead-us-to-over-eat-as-adults

Scientific Inquiry

Anwar, Y. (2018). Poor grades tied to class times that don't match our biological clocks. *Berkeley News.* Retrieved from http://news.berkeley.edu/2018/03/29/social-jetlag/

Chodosh, S. (2017). Cat poop parasites don't actually make you psychotic. *Popular Science.* Retrieved from https://www.popsci.com/cat-Toxoplasma-psychosis

Collins, C. (2016). Even insects have distinct personalities – research finds. *Horizon.* Retrieved from https://horizon-magazine.eu/article/even-insects-have-distinct-personalities-research-finds_en.html

Crawford, T. (2015). Music taste linked to brain type. *The Naked Scientists.* Retrieved from https://www.thenakedscientists.com/articles/science-news/music-taste-linked-brain-type

Dublin City University. (2018, January 19). Caffeine's sport performance advantage for infrequent tea and coffee drinkers. *ScienceDaily.* Retrieved June 20, 2018 from www.sciencedaily.com/releases/2018/01/180119090348.htm

Henion, A. (2018). Does dim light make us dumber? *MSU Today.* Retrieved from https://msutoday.msu.edu/news/2018/does-dim-light-make-us-dumber/

Kaiser, J. (2015). NIH to end all support for chimpanzee research. *Science.* Retrieved from http://www.sciencemag.org/news/2015/11/nih-end-all-support-chimpanzee-research

Maldarelli, C. (2016). Here's why twin studies are so important to science and NASA. *Popular Science.* Retrieved from https://www.popsci.com/heres-why-twin-studies-are-so-important-to-science-and-nasa

North Carolina State University. (2017, September 28). Driving speed affected when a driver's mind 'wanders'. *ScienceDaily.* Retrieved June 2, 2018 from www.sciencedaily.com/releases/2017/09/170928103057.htm

San Diego State University. (2018, January 22). Screen-addicted teens are unhappy: A new study finds that more screen time is coincides with less happiness in youths. *ScienceDaily*. Retrieved June 20, 2018 from www.sciencedaily.com/releases/2018/01/180122091249.htm

Sandle, T. (2018). Behavior in high school predicts life chances. *Digital Journal.* Retrieved from http://www.digitaljournal.com/life/lifestyle/behavior-in-high-school-predicts-life-chances/article/516422

Society of Interventional Radiology. (2018, March 21). Freezing hunger-signaling nerve may help ignite weight loss: Pilot study demonstrates safety of new weight-loss treatment for mild-to-moderate obesity. ScienceDaily. Retrieved June 20, 2018 from www.sciencedaily.com/releases/2018/03/180321090935.htm

Science News. (2018). Study: Bonobos prefer hinderers over helpers. Retrieved from http://www.sci-news.com/biology/bonobos-prefer-hinderers-05602.html

Stockholm University. (2018, February 27). Our reactions to odor reveal our political attitudes, survey suggests. *ScienceDaily.* Retrieved August 23, 2018 from www.sciencedaily.com/releases/2018/02/180227233253.htm

University of Oxford. (2017, December 13). Dinosaur parasites trapped in 100-million-year-old amber tell blood-sucking story: Amber containing tick grasping a dinosaur feather is first direct fossil evidence of ticks parasitizing dinosaurs. *ScienceDaily*. Retrieved June 20, 2018 from www.sciencedaily.com/releases/2017/12/171213104735.htm

University of Portsmouth. (2017, October 19). Dogs are more expressive when someone is looking. *ScienceDaily*. Retrieved June 20, 2018 from www.sciencedaily.com/releases/2017/10/171019100944.htm

University of York. (2018, January 16). No evidence to support link between violent video games and behavior. *ScienceDaily*. Retrieved June 20, 2018 from www.sciencedaily.com/releases/2018/01/180116131317.htm

Vanderbilt University. (2017, November 29). Sorry, Grumpy Cat: Study finds dogs are brainier than cats. *ScienceDaily*. Retrieved June 20, 2018 from www.sciencedaily.com/releases/2017/11/171129131341.htm

Study of Life

Agapakis, C. (2014). Which bacteria are in my poop? It depends where you look. *Scientific American.* Retrieved from https://blogs.scientificamerican.com/oscillator/which-bacteria-are-in-my-poop-it-depends-where-you-look/

Byrne, J. (2011). Peptidoglycan - The bacterial wonder wall. *Scientific American.* Retrieved from https://blogs.scientificamerican.com/disease-prone/peptidoglycan-the-bacterial-wonder-wall/

Chodosh, S. (2017). Swimming pools are full of poop, but they probably won't make you sick. *Popular Science.* Retrieved from https://www.popsci.com/swimming-pools-are-gross

Forsyth Institute. (2018, May 8). Cellular messengers communicate with bacteria in the mouth: Results of the study could lead to new treatments for diseases such as periodontitis. *ScienceDaily*. Retrieved June 20, 2018 from www.sciencedaily.com/releases/2018/05/180508131018.htm

Heid, M. (2018). How often should I clean my phone? *Time.* Retrieved from http://time.com/5310453/clean-my-phone/

King, A. (2018). "Mummies" tummies to reveal digestive evolution. *EAPS.* Retrieved from https://eapsweb.mit.edu/news/2018/mummies-tummies-reveal-digestive-evolution

Kupferschmidt, K. (2016). Researchers discovered that there's a microbial war happening in our nasal cavities. *Science.* Retrieved from http://www.sciencemag.org/news/2016/07/new-antibiotic-found-human-nose

Maldarelli, C. (2018). Probiotics might help your allergies, but we're still not sure how. *Popular Science.* Retrieved from https://www.popsci.com/can-probiotics-help-my-allergies

Miller, S. (2017). Rabbit, dog, human: How one bacterial infection spread. *Live Science.* Retrieved from https://www.livescience.com/60230-woman-got-rabbit-fever-from-dog.html

Milliken, G. (2015). Are viruses alive? New evidence says yes. *Popular Science.* Retrieved from https://www.popsci.com/new-evidence-that-viruses-are-alive

Reinberg. S. (2018). Kitchen towels laden with bacteria. *HealthDay.* Retrieved from https://consumer.healthday.com/infectious-disease-information-21/bacteria-960/kitchen-towels-laden-with-bacteria-734755.html

Stevens, A. (2016). Powered by poop and pee? *Science News for Students.* Retrieved from https://www.sciencenewsforstudents.org/article/powered-poop-and-pee

Telis, G. (2013). Got pimples? You may need better bacteria. *Science.* Retrieved from http://www.sciencemag.org/news/2013/02/got-pimples-you-may-need-better-bacteria

The Fecal Transplant Foundation. (n.d.) What is FMT? Retrieved from http://thefecaltransplantfoundation.org/what-is-fecal-transplant/

University of Toronto. (2017, December 22). A 508-million-year-old sea predator with a 'jackknife' head: Oldest close parent of spiders, scorpions and horseshoe crabs evolved sophisticated head to hunt and eat small shelly animals. *ScienceDaily*. Retrieved June 20, 2018 from www.sciencedaily.com/releases/2017/12/171222090334.htm

University of Oxford. (2018, January 25). The bacterial 'Game of Thrones': Understanding bacterial wargames inside our body. *ScienceDaily*. Retrieved June 20, 2018 from www.sciencedaily.com/releases/2018/01/180125135553.htm

Request for a Review

I sincerely hope you enjoyed using this educational tool as much as I enjoyed writing it. If you did, I would greatly appreciate a short review on Amazon or your favorite book or instructional materials website. Reviews are crucial for letting others know of the value of any instructional resource, and even just a line or two can make a huge difference.

Connect with The Author

G. Katz Chronicle has over three decades of teaching experience in primary, secondary, and higher education. Having taught all major content areas in primary and middle school, as well as education theory and biology at the secondary and college level, she most enjoys teaching the Life Sciences.

With a PhD in Curriculum and Instruction, she, too, is a perpetual student of traditional education and all areas of self-improvement.

G. Katz is an abbreviated form of Gertrude Katz. Gertrude maintains a Biology Teaching Blog where she provides her general instructional plans, reasoning, and thoughts for the NY State Living Environment Course. The blog consists of ten posts, which correspond to ten instructional units, and include unit presentations and videos. Gertrude invites you to check it out at: GertrudeKatzChronicles.com. Connect with her on other social media, as well!

Made in the
USA
Lexington, KY